SpringerBriefs in Molecular Science

Chemistry of Foods

Series Editor

Salvatore Parisi, Al-Balqa Applied University, Al-Salt, Jordan

The series Springer Briefs in Molecular Science: Chemistry of Foods presents compact topical volumes in the area of food chemistry. The series has a clear focus on the chemistry and chemical aspects of foods, topics such as the physics or biology of foods are not part of its scope. The Briefs volumes in the series aim at presenting chemical background information or an introduction and clear-cut overview on the chemistry related to specific topics in this area. Typical topics thus include:

- Compound classes in foods—their chemistry and properties with respect to the foods (e.g. sugars, proteins, fats, minerals, …)
- Contaminants and additives in foods—their chemistry and chemical transformations
- Chemical analysis and monitoring of foods
- Chemical transformations in foods, evolution and alterations of chemicals in foods, interactions between food and its packaging materials, chemical aspects of the food production processes
- Chemistry and the food industry—from safety protocols to modern food production

The treated subjects will particularly appeal to professionals and researchers concerned with food chemistry. Many volume topics address professionals and current problems in the food industry, but will also be interesting for readers generally concerned with the chemistry of foods. With the unique format and character of SpringerBriefs (50 to 125 pages), the volumes are compact and easily digestible. Briefs allow authors to present their ideas and readers to absorb them with minimal time investment. Briefs will be published as part of Springer's eBook collection, with millions of users worldwide. In addition, Briefs will be available for individual print and electronic purchase. Briefs are characterized by fast, global electronic dissemination, standard publishing contracts, easy-to-use manuscript preparation and formatting guidelines, and expedited production schedules.

Both solicited and unsolicited manuscripts focusing on food chemistry are considered for publication in this series. Submitted manuscripts will be reviewed and decided by the series editor, Prof. Dr. Salvatore Parisi.

To submit a proposal or request further information, please contact Dr. Sofia Costa, Publishing Editor, via sofia.costa@springer.com or Prof. Dr. Salvatore Parisi, Book Series Editor, via drparisi@inwind.it or drsalparisi5@gmail.com

More information about this subseries at http://www.springer.com/series/11853

Moawiya A. Haddad · Mohammed I. Yamani ·
Da'san M. M. Jaradat · Maher Obeidat ·
Saeid M. Abu-Romman · Salvatore Parisi

Food Traceability in Jordan

Current Perspectives

Springer

Moawiya A. Haddad
Department of Nutrition and Food
Processing
Faculty of Agricultural Technology
Al-Balqa Applied University
Al-Salt, Jordan

Da'san M. M. Jaradat
Department of Chemistry
Al-Balqa Applied University
Al-Salt, Jordan

Saeid M. Abu-Romman
Department of Agricultural Biotechnology
Faculty of Agricultural Technology
Al-Balqa Applied University
Al-Salt, Jordan

Mohammed I. Yamani
Department of Nutrition and Food
Technology
Faculty of Agriculture
University of Jordan
Amman, Jordan

Maher Obeidat
Department of Biomedical Analysis
Faculty of Science
Al-Balqa Applied University
Al-Salt, Jordan

Salvatore Parisi
Department of Nutrition and Food
Processing
Faculty of Agricultural Technology
Al-Balqa Applied University
Al-Salt, Jordan

ISSN 2191-5407   ISSN 2191-5415 (electronic)
SpringerBriefs in Molecular Science
ISSN 2199-689X   ISSN 2199-7209 (electronic)
Chemistry of Foods
ISBN 978-3-030-66819-8   ISBN 978-3-030-66820-4 (eBook)
https://doi.org/10.1007/978-3-030-66820-4

This Springer imprint is published by the registered company Springer Nature Switzerland AG
The registered company address is: Gewerbestrasse 11, 6330 Cham, Switzerland

Contents

Chapter 1
An Introduction to Food Traceability

Abstract This chapter concerns the evolution of food traceability matters in the current market of traditional foods and beverages. At present, traceability is only one of the many requirements food industries are forced to comply with. These challenges are: microbiological failures affecting food safety; chemical and physical contaminants into food products; other product-related (intrinsic) menaces against food safety, in terms of consumers' health; the demonstrable evidence of risk assessment in terms of clear and reliable documentation concerning safety, integrity, and legal designation of food and beverage products; and the evidence of continuous improvement by means of clear standard operative procedures, good manufacturing practices, and the execution of corrective/preventive actions against unavoidable food-related failures. The intrinsic connection between 'evidence' or 'demonstration' on the one side and the existence of documentations able to trace the production of foods and beverages on the other side should be established. This topic can be discussed by different viewpoints: the regulatory angle; technological perspectives; mathematical theories (networks, hubs, and nodes); and the opinion of food consumers. In addition, traceable food products may be also an interesting legacy for many geographical and ethnic cultures.

Keywords Hub · Food Business Operator · Network · Node · Off-line · Food Packaging Operator · Traceability

Abbreviations

AD/N	Average degree for node
GSFA	Codex General Standard of Food Additives
Comp	Complexity
EU	European Union
FAO	Food Agriculture Organization of the United Nations
FBO	Food Business Operator
F&B	Food and beverage
FMRIC	Food Marketing Research and Information Center

© The Author(s), under exclusive license to Springer Nature Switzerland AG 2021
M. A. Haddad et al., *Food Traceability in Jordan*,
Chemistry of Foods, https://doi.org/10.1007/978-3-030-66820-4_1

1

FPO   Food Packaging Operator
IFT   Institute of Food Technologists
In   Interconnection
MD/N   Maximum degree per node
N   Node
OL   Off-Line

1.1 A General Introduction to Food Traceability

The modern industry of foods and beverages is forced to face many problems and concerns when speaking of public hygiene, food safety, regulatory requirements, traceability and authenticity questions, etc. (Beulens et al. 2005; Delgado et al. 2016, 2017; Mania et al. 2016a, b-2018; Parisi 2002, 2016; Parisi et al. 2016; Perreten et al. 1997; Phillips 2003; Pisanello 2014; Zanoli and Naspetti 2002; Zhang 2015). These matters could be mentioned in a general way as follows, considering that the list is not exhaustive:

1. Microbiological risks in terms of food safety and commercial requirements (Sheenan 2007a, b; Silva and Malcata 2000; Steinka and Parisi 2006);
2. Chemical risks affecting food safety, including decomposition, hydrolysis, and shelf-life determination (Barbieri et al. 2014; Vairo Cavalli et al. 2008)
3. Detection of foreign substances in foods (metals, plastics materials, wooden materials, nano-chemicals, etc.);
4. Other food-related risks affecting consumers' health;
5. Demonstrable evidence of risk assessment (safety, integrity, presence and decla-ration of food additives, nutritional claims, and legal designation of food and beverage products) (GSFA 2017; Haddad et al. 2020a, b; Silvis et al. 2017);
6. Reliable evidence of continuous improvement (in terms of standard opera-tive procedures, good manufacturing practices, etc.), also with concern to food packaging requirements (Italian Institute of Packaging 2011).

Actually, the problem of demonstrable dangers (microbiological contaminations; the detection of prohibited food additives or foreign substances) can be understood by the main part of consumers, while other risks such as traceability may be difficultly managed and explained to the general population (Golan et al. 2004a, b; Mania et al. 2016a, b, 2018), depending on the 'active player' or 'subject' (Lauge et al. 2008; Mitroff et al. 1987; Olsen and Borit 2018; Parisi 2016; Haddad and Parisi 2020a, b; Parisi 2019).

By a general viewpoint, the 'traceability' of foods and beverages might be consid-ered as the intrinsic connection between 'evidence' on the one side and the existence of documentations able to trace the production of foods and beverages on the other side (Parisi 2020). The importance of traceability is also notable when speaking of the reaction of food and beverage consumers (Buhr 2003; Cheek 2006; Chrysochou et al. 2009; Giraud and Halawany 2006; Golan et al. 2004a, b; Jasnen-Vullers et al. 2003; Sodano and Verneau 2004; Starbird and Amanor-Boadu 2006; Verbeke et al. 2007).

1.2   Some Definitions of Food Traceability. an International Perspective

It has been affirmed that traceability could be defined differently if the interested players/stakeholders are Food Business Operators (FBO), Official Inspection Bodies, Policy makers, or Food Consumers (Parisi 2020). Three definitions can be provided here with the aim of explaining better this concept.

According to the Food and Agriculture Organization of the United Nations (FAO) and its 'Food Traceability Guidance', food traceability is *'the ability to discern, identify and follow the movement of a food or substance intended to be or expected to be incorporated into a food, through all stages of production, processing and distribution'* (FAO 2017). Other official documentations are available on the National and the International level:

a. The 'Seafood Traceability Glossary: a guide to terms, technologies, and topics' document (Future of Fish, FishWise, and Global Food Traceability Center 2018);
b. The 'Handbook for Introduction of Food Traceability Systems (Guidelines for Food Traceability)' (FMRIC 2008);
c. The Food Safety Guidebook by Alberta Agriculture and Rural Development, Canada, 'Glossary of Food Safety Related Terms'—Appendix A (Alberta Agriculture and Rural Development 2014);
d. The 'Global Food Traceability Center, Glossary of Terms', Institute of Food Technologists (IFT), Chicago (IFT 2018);
e. The Regulation (EC) No 178/2002 of the European Parliament and of the Council (European Parliament and Council 2002).

The last document provides an interesting definition concerning traceability: *'The ability to trace and follow a food, feed, food producing animal or substance intended to be, or expected to be used for these products at all of the stages of production, processing and distribution'* (European Parliament and Council 2002).

Anyway, the practical basis of traceability cannot be fully understood without the comprehension of traceability pillars, as briefly expressed in Table 1.1, and some common points among different traceability definitions, as shown in Table 1.2. This matter is of critical importance in the modern world of foods, beverages, and food-contact materials worldwide (Epelbaum and Martinez 2014; Fiorino et al. 2019; Herzallah 2012; Hoorfar et al. 2011; Kim et al. 2018; Meuwissen et al. 2003; Schwägele 2005), because many controls and evaluations have to be based on reliable information which have to (a) be stored as soon as possible and (b) made available on request to interested 'stakeholders' with strict (and reasonable) timelines.

Table 1.1 Traceability pillars according to several documents and Organizations (Alberta Agriculture and Rural Development 2014; FAO 2017; FMRIC 2008; Future of Fish, FishWise, and Global Food Traceability Center 2018; IFT 2018; European Parliament and of the Council 2002; Parisi 2020)

Traceability pillars	Explanation
Batch Traceability	These words concern the basic work of a management system able to find and give evidence of information concerning moveable units under the same lot or batch number. The simple lot identification concerns similar features such as harvesting dates, peculiar names, and so on
Commercial Traceability	Similar to batch traceability, without the mention and public sharing of proprietary and/or confidential information
Electronic traceability	These words concern the amount of information obtained by means of a traceability management system based only on paperless information
External traceability	The basic work of a management system able to find and give evidence of information concerning moveable units under the same lot or batch number, as they move outside of the facility of the traceability manager
Internal traceability	The basic work of a management system able to find and give evidence of information concerning moveable units under the same lot or batch number, as they move into the facility of the traceability manager
Paper-based traceability	Differently from electronic traceability, these words concern the amount of information obtained by means of a traceability management system based only on paper-based documents information (although scanned images may be stored and used)
Chain traceability	The sum of all traceability information along the whole food chain, without exclusions (all FBO are considered)
One-step-back traceability	The sum of all important traceability information concerning received units (the information has to be received by the supplier)
One-step-forward traceability	The sum of all important traceability information concerning sold units (the buyer has to be clearly identified)

1.3   Food Traceability: Stakeholders, Hubs, and Nodes

Food traceability procedures have to concern all food and beverage (F&B) items and related players (stakeholders) at the same time. In this ambit, F&B stakeholders should be (Alberta Agriculture and Rural Development 2014; Brunazzi et al. 2014; FAO 2017; FMRIC 2008; Future of Fish, FishWise, and Global Food Traceability Center 2018; Haddad and Parisi 2020a, b; Mania et al. 2016b; Parisi 2012a, 2013, 2016, 2020):

1. Food business operators (FBO) which are involved in production. This category includes primary processors (growers, etc.), produce packers and re-packers,

Table 1.2 Common features of many traceability viewpoint (Alberta Agriculture and Rural Development 2014; FAO 2017; FMRIC 2008; Future of Fish, FishWise, and Global Food Traceability Center 2018; Hosch and Blaha 2017; IFT 2018; European Parliament and of the Council 2002; Mania et al. 2018; Parisi 2020)

Traceability Common Points	Explanation
Identification	Each material used for foods and beverages, or the food and beverage itself, is identified by means of data. One (or more) dataset represents virtually the unit
Ability	It should be requested to all possible players of the food chain because of the nature of the material identified
Movement	Each material used for foods and beverages, or the food and beverage itself, can be identified by means of data on condition that an input is considered at the start of a process, and an output is generated at the end
Downstream	The direction of the supply chain with all possible stages in the food production process involving processing, packaging, and distribution
Upstream	The opposite direction of downstream in the supply chain
'Tracing forward' or 'tracking'	The information concerning moveable units and the related process(es) can be considered in the downstream direction (from raw materials to final products) or in the upstream direction (from final products to raw materials)
'Tracking back' or 'tracing'	The information concerning moveable units and the related process(es) in the upstream direction (from final products to raw materials)
Lot or Batch	A food unit, or a homogeneous group of food units, have to be clearly identified by means of reliable identification keys such as production dates, best-of-use dates, sequential numbers, etc.
Large-scale (or external) traceability	The traceability system can operate with the synergic action of many operators
Small-scale traceability	The system should operate only within the boundaries of an operator. Also defined 'internal traceability' (Table 1.1)

    livestock producers, food processors, suppliers of crop protection and/or seeds and plants, etc.

2. F&B packaging producers, also named Food Packaging Operators (FPO);
3. F&B traders, distributors, importers, and exporters. Third-party logistics service providers may be included here;
4. F&B mass retailers (generally working on a large-scale dimension, although little and average-sized retail stores are often found in this sector);
5. Other FBO: catering operators, chain restaurants, universities, hotels, government and hospital cafeterias, etc.
6. F&B consumers;
7. Official Safety Authorities;

8. The Academia, third-party certification bodies, food lawyers, technical consul-
    tants, journals, magazines, private research centres, social media, etc.

The network of interested players should comprehend only 'inner' stakeholders
in the strict ambit of food production/movement/processing/distribution, as shown
in Fig. 1.1. This basic structure comprehends five possible connections or 'nodes'
representing five players in the food chain. Interestingly, the ideal process should
follow a simple direction, from the left to the right only, and each theoretical F&B
unit (or item) would be obliged to enter a specific node, from primary production to
final retail, without 'skipping' options (Parisi 2020).

The 'node' term comes from 'Seafood Traceability Glossary: a guide to terms,
technologies, and topics' document (Future of Fish, FishWise, and Global Food
Traceability Center 2018). In brief, each node can be considered as '*a distinct entity
in a supply chain (...) that may be responsible for capturing, inputting, storing,
or sharing data*'. As a result, a food traceability network—similarly to the real
commercial network of FBO and other players—would be an interconnected matrix
of nodes (Parisi 2020), from harvesting activities or similar options, towards the
market and/or processors, distributors, and retailers.

This representation is simple enough. Unfortunately, the real food network could
have more than a single FBO or stakeholder per step. In other time, F&B units
have to be defined as a single entity, while interested players may be more than one
operator only per step. Usually, the main part of food products appears to be sold
by retailers, in the European Union (EU) at least (Fig. 1.2, pathways B and C). On
the other hand, many primary producers and processors could sell their products to

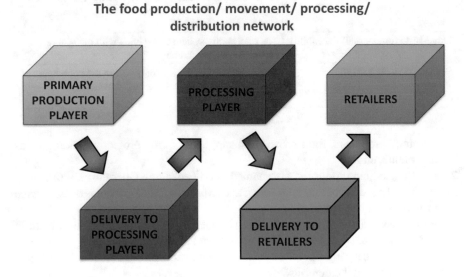

Fig. 1.1 This basic structure comprehends five possible connections or 'nodes' representing five
players in the food chain. F&B items should not skip mentioned steps (Parisi 2020)

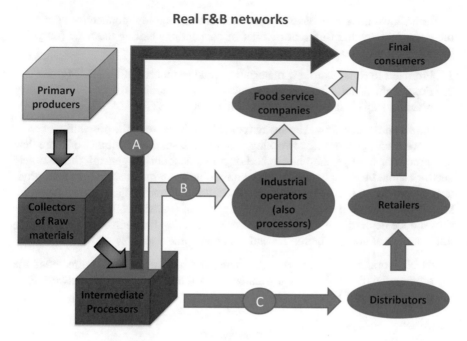

Fig. 1.2 Real F&B networks could have more than a single FBO or stakeholder per step. Many primary producers and processors could address their products to food service companies (catering, etc.) and final consumers at the same way (pathway A, red colour), or to industrial operators (food companies where cheeses are only one of the many ingredients; pathway B, yellow colour), or to distributors (pathways C, orange colour)

non-retailer companies for further processing (pathway A). This matrix is shown in Fig. 1.2: the complexity of similar networks depends also on two features (Future of Fish, FishWise, and Global Food Traceability Center 2018; Parisi 2020):

a. The 'fragmentation' (certain steps are located in different places or by different companies at the same times);
b. The 'horizontal concentration' of nodes in the food chain (only a small number of companies operate at a single step, accumulating food units from a multiplicity of small suppliers). In this situation, the flow of commodities and the price of F&B products is influenced and substantially managed into a single node.

The study of food traceability networks could be simplified by means of the analysis of the degree of different nodes. In ould be considered with attention (Biggs et al. 1986; Chartrand 1985; Nykamp 2018; Parisi 2020):

1. The degree of a node may be defined as the number of interconnections between a specified node and surrounding nodes
2. The degree distribution has to be defined as the probability distribution of all possible degrees in relation to the entire network.

Each shown node may have an exact number of 'ingoing' connections (entering the node) and another different number of connections exiting the node (outgoing the node). In addition, there are two possible directions:

1. From left to right, from raw materials to final products (UPSTREAM), and
2. From right to left, from final products (or intermediates) to the immediately previous step (in the opposite direction, or DOWNSTREAM).

The second option should be always considered because of the possibility of recycling operations concerning reworking materials. Also, the possibility of more than one processor only per step has to be taken into account. In general, the complete network should have only a preferential direction: from raw materials to final products, step by step. In this situation, the distribution of node degrees could be represented by means of a simple binomial distribution (Parisi 2020) where the X-axis shows node degrees and the Y-axis concerns the ration between each node and the total number of nodes. Figures 1.3 and 1.4 show three options:

a. #1 Network: a four-node network where each interconnection is directed in the preferential direction only (raw materials → final products). This network is on the upper side of Fig. 1.3.

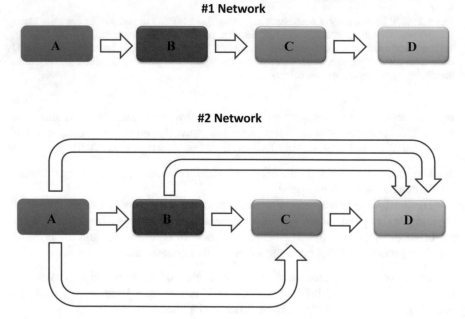

Fig. 1.3 Two simple supply chain networks with four nodes A, B, C, and D. nodes. Figures 1.3 and 1.4 show three options: (d) #1 Network: a four-node network where each interconnection is directed in the preferential direction only (raw materials → final products). This network is on the upper side of Fig. 1.3 (e) #2 Network: a four-nodes network showing three specific nodes with three, two, and one interconnections in the preferential direction (raw materials → final products), respectively. This model is on the lower side of Fig. 1.3)

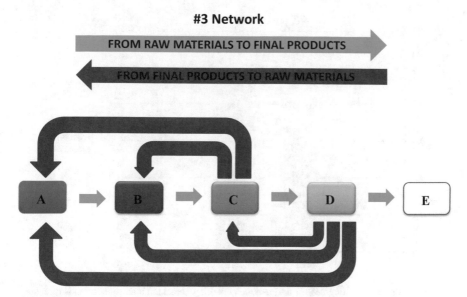

Fig. 1.4 A five-nodes network showing four nodes with four interconnections, but nodes C and D have also two outgoing connections directed towards A & B, and B & C, respectively

b. #2 Network: a four-nodes network showing three specific nodes with three, two, and one interconnections in the preferential direction (raw materials → final products), respectively. This model is on the lower side of Fig. 1.3.

c. #3 Network: a five-nodes network (Fig. 1.4) showing four nodes with four inter-connections, but nodes C and D have also two outgoing connections directed towards A & B, and B & C, respectively.

Each network can be expressed by means of a distribution showing node degrees on the X-axis, and the fraction of nodes on the Y-axis (Parisi 2020). Figures 1.5, 1.6, and 1.7 show the situation concerning #1, #2, and #3 networks, respectively.

The existence of more than two connections per node (one ingoing and one outgoing relationship) means that certain nodes can represent a 'hub' (high number of interconnections). Consequently, the distribution of ingoing and outgoing relations in two directions has to be graphed two times with relation to in-degree and out-degree on the X-axis.

In general, the distribution of outgoing relations in the non-preferential (DOWN-STREAM: final products → raw materials) direction is low enough when speaking of food chains. For this reason, the complexity of these networks (including their virtual representation, traceability systems) is directly dependent on the distribution of ingoing relations in the preferential direction (Nykamp 2018; Parisi 2020). The higher the ratio between the total number of degrees and the number of nodes, the more complex the whole network, and vice versa. Moreover, three considerations should be evaluated in this specific ambit (Parisi 2020):

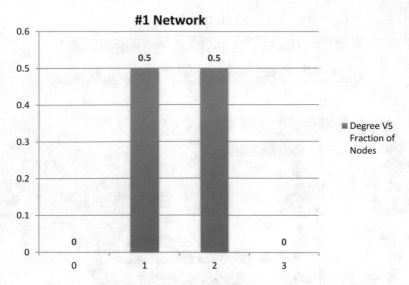

Fig. 1.5 Distribution of degrees for a four-node network where each interconnection is directed in one direction only (from raw materials to final products, UPSTREAM). The #1 network is on the upper side of Fig. 1.3

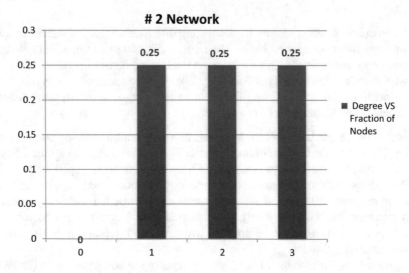

Fig. 1.6 Distribution of degrees for a four-node network shows three specific nodes with three, two, and one interconnections in the preferential direction (raw materials → final products, UPSTREAM), respectively. The #2 network is on the lower side of Fig. 1.3

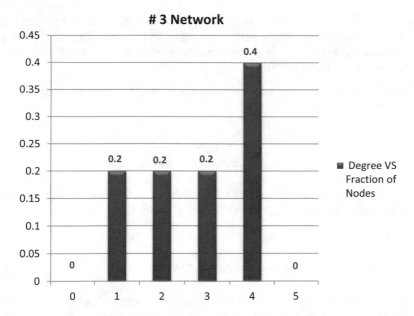

Fig. 1.7 Distribution of degrees for a five-node network (Fig. 1.4) shows three nodes with four interconnections, but node C has two outgoing connections directed towards A and B (opposite direction if compared with raw materials to the final product)

a. A distribution with low ratios tends to be placed on the left of the graph. The network tends to be simple with a high number of nodes showing low degrees. Consequently, this model would be easily managed provided that each node can perform well alone, but this situation is not simple (and a general manager with monitoring activity for the whole network would be needed…)
b. A Gaussian or quasi-Gaussian distribution should be placed preferentially at the centre of the graph (average or high ratios). The system tends to be more and more complex with an average number of nodes (hubs) having high degrees. The network might be managed by one or more of these hubs; on the other hand, the global performance of similar systems depends entirely on the performance of these hubs only (different companies). #1 and #2 Networks appear in this situation (Figs. 1.5 and 1.6)
c. A non-Gaussian distribution showing only high ratios (a very small number of hubs). In this ambit, the management of the whole information (or logistic) system should be managed by one of these hubs only. The risk of a complete downfall is linked to the ability of this hub (or company). #3 network might be similar, while Fig. 1.7 shows a complex system with these features (number of interconnections: 18). The related distribution with high ratios tends to be placed on the right side of the graph.

In the ambit of traceability and food chain networks, the ratio between the total number of degrees and the number of nodes may show related complexity if each node has 0, 1, or 2 maximum connections. Also, the comparison between this ratio and the same value obtained with the maximum degree of nodes would give a percentage idea of the complexity. By a mathematical point of view, the following procedure may be proposed (Parisi 2020):

a. The ratio between interconnections (In) and nodes (N), named 'average degree' for node (AD/N) is:

$$AD/N = In/N \times 100 \tag{1.1}$$

b. The 'complexity' (Comp) of the whole network is the % ratio between AD/N and the maximum degree per node (MD/N), as follows:

$$Comp = AD/N / MD/N \times 100 \tag{1.2}$$

Consequently, should In $= 6$ and $N = 4$ only for a network while MD/N is 2.0, the following results could be obtained by means of Eqs. 1.1 and 1.2:

$$AD/N = 6/4 = 1.5,$$
$$and$$
$$Comp = 1.5/2.0 \times 100 = 75\%$$

On the other hand, should In $= 14$ and $N = 5$ with MD/$N = 4$ (this high value means that one of nodes at least is a hub), the following results could be obtained by means of Eqs. 1.1 and 1.2:

$$AD/N = 14/4 = 2.8,$$
$$and$$
$$Comp = 2.8/4.0 \times 100 = 70\%$$

Two reflections should be made on these bases. The first of these examples concerns a simple network with no hubs (AD/$N = 2$) and a notable complexity (75%). This situation is explainable because AD/N and MD/N are close enough. On the other hand, the second example shows MD/$N > 2$ (one or more hubs), as also confirmed by calculated AD/N (2.8 > 2). However, complexity seems lower than in the first example (70 vs. 75%). The difference between the two situations is the recommended use of a centralised managed in the second network because one hub at least is present, and it could represent the possible 'weak ring' of the chain. Should one (or more) of these nodes fail to transfer information, the whole traceability system would easily collapse (Parisi 2020). On the other side, the first system would not necessarily need a centralised manager. On these bases, the performance of these networks would be similar. Anyways, the important factor is the MD/N value: should this number be 5.0 in the second situation, Comp value would be 56% (lower complexity), but the hub with node degree = 5.0 would be the real and critical point of the system… Should all food chain players agree, this step—and the interested players—would represent the real monitoring manager avoiding the possible downfall (Parisi 2020).

1.4 The Food Unit as the Basic Pillar of Food Traceability

Each food item, also named 'food unit', has to be clearly identified. In detail, several information concerning the identification of food units should be provided (FAO 2017; Parisi 2020):

a. Name of the FBO (contact addresses, phone, e-mail addresses, other company data, etc., concerning both the main headquarter and all possible additional sites, if needed)
b. The name of the person working on traceability services for the FBO (it is strongly recommended that this manager has to be always available, unless one or more substitutes are present and specifically nominated)
c. The management plan for traceability of the FBO. This plan may be part of general quality management systems in the company.
d. The shelf life or expiration date of the food unit
e. Recommended storage procedures for the food unit, including logistic aspects
f. The Nation of origin, with additional data concerning possible outsourcers, external manufacturers, and/or exporter companies.

All information have to be adequately recorded and stored for each product, food, or substance intended to be found into foods, or used in food production and/or packaging (Piergiovanni and Limbo 2016). Consequently, there are only traceable 'food units'. According to the FAO (2017), three food categories for sale may be considered:

1. A packaged food or beverage unit (also constituted of more than one unit in a single macro-product unit)
2. A logistic unit (sum of many units). In this ambit, the final consumer is not accustomed to see logistic units such as prepared mass of ordered commodities placed on wooden or plastic pallets (Steinka and Parisi 2006)
3. A shipment of F&B units.

The problem of traceability is linked to two different aspects (Parisi 2020):

a. F&B units have to be traceable and traced always (in each moment)
b. Traceability records, managed by centralised monitoring agents or by each player of the traceability network (Sect. 1.3), should be made available on request immediately, and stored for a specified time.

With reference to the last point, it is difficult to find a general quality manager which is able to monitor also traceability. In other words, a different function—the traceability manager—should be considered.

1.5 The Flow of Input and Output Information with Mathematical Implications

According to the 'Hygiene, Integrity, Traceability, and Sharing' strategy (Haddad and Parisi 2020a, b), traceability procedures are critical and represent one of main pillars in the quality and safety management of food and beverage industries. This requisite is mandatory, required worldwide (Allata et al. 2017; Chen 2017; King et al. 2017; Lewis et al. 2016), and it is mentioned also in certification standards such as Global Standard for Food Safety (by the British Retail Consortium, UK), and the International Featured Standard (IFS) Food (Bitzios et al. 2017; Jin et al. 2017; Mania et al. 2018; Nicolae et al. 2017; Parisi 2020; Stilo et al. 2009; Telesetsky 2017). The list of interested F&B products and services is virtually infinite (Kok 2017; Lacorn et al. 2018; Moyer et al. 2017; Pisanello and Caruso 2018; Silvis et al. 2017).

With concern to mandatory requirements, the European Regulation (EC) No 178/2002 is a good explanation when speaking of the European ambit, even if this Regulation is completed with other specific regulator norms (European Commission 2004, 2006, 2008, 2011, 2013; European Parliament and Council 2002, 2003a, b, 2009, 2014).

However, the problem of traceability is also a matter of mathematical balances… because all ingredients of interest have to be inserted into a 'flow' towards the final product. This concept is true from the production and technological viewpoint. At the same time, the flow of information is a virtual representation of the technological process, and the whole traceability system (as a mass of organised paper documents, or an electronic storing and recording software) has to take this point into account. Figure 1.8 shows a general figure while five different ingredients (and related information) enter the flow of input data towards the final food or beverage (ideally representing the output product with associated output data: name, brand, lot, etc.).

A premise should be made in the world of F&B products concerning the nature of entering information. What about… water? Water—sometimes named 'the blue gold'—is absolutely needed in food industries for many purposes… including the role of ingredient (and solvent also…). Basically, the equation representing the balance between ingredients and the final product:

$$[\text{Sum of all ingredients}] = [\text{Finished product}] \qquad (1.3)$$

should contain a mistake: the absence of water, with several exceptions… Consequently, this balance may be replaced with the following Eq. 1.4, when needed (Mania

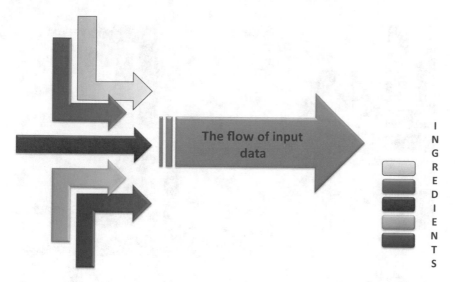

Fig. 1.8 Five different ingredients (and related information) enter the flow of input data towards the final food or beverage (ideally representing the output product with associated output data: name, brand, lot, etc.)

et al. 2018):

$$\left[\text{Sum of all ingredients excluding added water}\right] + \left[\text{Added water}\right]$$
$$= \left[\text{Finished product}\right] \qquad (1.4)$$

Another important consideration should concern the number of 'final products', because normal processes generate the final product itself and 'n' possible by-products. As a result, Eq. 1.4 should be amended again. These by-products may be defined 'Reworking' or 'Off-Line' (OL) products, corresponding to the difference between the total amount of produced intermediates in a specific process, and the quantity of final products that can be sold. Also, added water could be partially lost (and other ingredients could loss water...). Consequently (Mania et al. 2018; Parisi 2020):

$$\left[\text{OL}\right] = \left[\text{Sum of products}\right] - \left[\text{Marketable products}\right] \qquad (1.5)$$

Finally, Eq. 1.4 would be amended correctly as follows:

$$\left[\text{Sum of all ingredients excluding added water}\right] + \left[\text{Added water}\right]$$
$$= \left[\text{Finished product}\right] + \left[\text{OL}\right] + \left[\text{Lost water}\right] \qquad (1.6)$$

where:

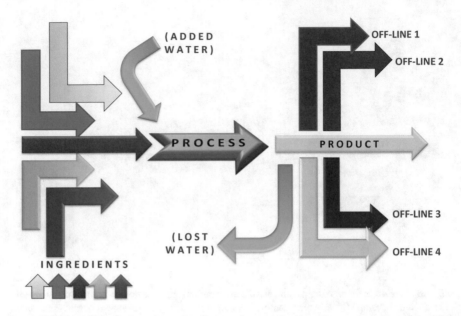

Fig. 1.9 Traceability requires all possible input and output data in a process, including also added water and lost water (Eq. 1.6). This approach should give evidence of all raw materials and other ingredients (on the left), while output information is on the right (marketable products, OL, and possible lost water). This example shows five ingredients and added water on the left side of the image, and the final product with four OL and lost water on the right

$$[\text{Finished product}] = [\text{marketable products}] \qquad (1.7)$$

The situation expressed by Eq. 1.6 is generally observed in the F&B industry, although the application depends on the peculiar product and the concerned industries… Fig. 1.9 shows the flow of input (on the left = ingredients) and output (on the right = final products + OL + possible lost water) information concerning the flow of traceability data.

1.6 Food Traceability… and Packaging Materials?

Obligations and requirements linked to food traceability should consider also the role of food-contact materials, probably seen by the main part of interested stakeholders as 'accessory' ingredients (Mania et al. 2016a, b, 2018; Parisi 2009). However, the role of these non-edible materials has been recently considered with extreme attention, in accordance with the (EC) Regulation No 1935/2004 in this ambit (European Parliament and Council 2004). This regulatory document of the European Union states clearly (Article 17, point 1) that '*The traceability of materials and articles shall be ensured at all stages in order to facilitate control, the recall of defective products, consumer information and the attribution of responsibility*. As a result, FBO have to assure the correct identification of food packaging materials (FCM) when speaking of the following data (Parisi 2020):

1. Origin;
2. Name and identification of the initial supplier;
3. Date of production (also named batch);
4. Name of the consignee;
5. And other data of interest, considering the non-edible nature of these materials.

Anyway, the meaning of 'traceability' in this ambit is not different from food-related definitions. In accordance with the (EC) Regulation No 1935/2004, Article 2, point 1, comma a, *'the ability to trace and follow a material or article through all stages of manufacture, processing and distribution'*.

As an obvious consequence, the extended traceability system should not only consider edible ingredients and products (Eq. 1.6, Fig. 1.9), but also food-contact materials at the same time! Also, the mandatory requisite of traceability (also considered by dedicated food certification systems) has to be satisfied by FBO and FPO at the same time and worldwide (Allata et al. 2017; Chen 2017; Kawecka 2014; King et al. 2017; Lewis et al. 2016; Mania et al. 2018; Parisi 2020; Stilo et al. 2009). Relations between food traceability and safety matters concern all components of F&B products, including FCM, and the possible list of reasons (including potential safety risks and commercial failures) could be really infinite (Kok 2017; Lacorn et al. 2018; Moyer et al. 2017; Parisi 2012a, b, 2013; Pisanello and Caruso 2018; Silvis et al. 2017).

For these reasons, it should be highlighted that traceability represents a great effort, aiming at giving the evidence of an entire flow of information among all interested players and stakeholders of the food chain with a continuous data exchanging (Mania et al. 2018). In this ambit, FBO and PBO should be able to demonstrate their compliance to traceability requirements always... This objective can be really difficult because of the virtually infinite list of F&B products on the national and international markets. This book would give a good analysis of traceability problems taking into account many information related to three categories of foods currently found in the Middle East, and particularly in Jordan. In the ambit of traceability and related data, Chapter 2 is dedicated to traditional *hummus* and related variations, while Chapters 3 and 4 concern Jordanian concentrated strained yogurt (*labaneh*) and dried fermented dairy foods (*jameed*).

References

Alberta Agriculture and Rural Development (2014) Food safety guidebook, Version 1.1, 2014, 7 November. Alberta Agriculture and Rural Development, Edmonton, Canada. Available https:// open.alberta.ca/publications/food-safety-guidebook. Accessed 29 Sept 2020

Allata S, Valero A, Benhadja L (2017) Implementation of traceability and food safety systems (HACCP) under the ISO 22000:2005 standard in North Africa: the case study of an ice cream company in Algeria. Food Control 79:239–253. https://doi.org/10.1016/j.foodcont.2017.04.002

Barbieri G, Barone C, Bhagat A, Caruso G, Conley Z, Parisi S (2014) The prediction of shelf life values in function of the chemical composition in soft cheeses. In: The influence of chemistry on new foods and traditional products. Springer International Publishing, Cham. https://doi.org/10. 1007/978-3-319-11358-6_2

Beulens AJM, Broens DF, Folstar P, Hofstede GJ (2005) Food safety and transparency in food chains and networks. relationships and challenges. Food Control 16(6):481–486. https://doi.org/ 10.1016/j.foodcont.2003.10.010

Biggs N, Lloyd EH, Wilson RJ (1986) Graph theory. Clarendon Press, Oxford, pp 1736–1936

Bitzios M, Jack L, Krzyzaniak SC, Xu M (2017) Dissonance in the food traceability regulatory environment and food fraud. In: Martino G, Karantininis K, Pascucci S, Dries L, Codron JM (eds) It's a jungle out there—the strange animals of economic organization in agri-food value chains. Wageningen Academic Publishers, Wageningen. https://doi.org/10.3920/978-90-8686-844-5_0

Brunazzi G, Parisi S, Pereno A (2014) The importance of packaging design for the chemistry of food products. Springer International Publishing, Cham. https://doi.org/10.1007/978-3-319-084 52-7

Buhr BL (2003) Traceability and information technology in the meat supply chain implications for firm organization and market structure. J Food Distrib Res 34(3):13–26. https://doi.org/10. 22004/ag.econ.27057

Chartrand G (1985) Introductory graph theory. Dover, New York

Cheek P (2006) Factors impacting the acceptance of traceability in the food supply chain in the United States of America. Rev Sci Tech Off Int Epiz 25(1):313–319

Chen PY (2017) A decision-making model for deterring food vendors from selling harmless low-quality foods as high-quality foods to consumers. J Food Qual Article ID 7807292, 1–8. https:// doi.org/10.1155/2017/7807292

Chrysochou P, Chryssochoidis G, Kehagia O (2009) Traceability information carriers. The technology backgrounds and consumers' perceptions of the technological solutions. Appet 53(3):322–331. https://doi.org/10.1016/j.appet.2009.07.011

Delgado AM, Almeida MDV, Parisi S (2017) Chemistry of the Mediterranean diet. Springer International Publishing, Cham. https://doi.org/10.1007/978-3-319-29370-7

Delgado AM, Vaz de Almeida MD, Barone C, Parisi S (2016) Leguminosas na Dieta Mediter-rânica—Nutrição, Segurança, Sustentabilidade. CISA - VIII Conferência de Inovação e Segurança Alimentar, ESTM- IPLeiria, Peniche, Portugal

Epelbaum FMB, Martinez MG (2014) The technological evolution of food traceability systems and their impact on firm sustainable performance: A RBV approach. Int J Prod Econ 150:215–224. https://doi.org/10.1016/j.ijpe.2014.01.007

European Commission (2004) Commission Regulation (EC) No 2230/2004 of 23 December 2004 laying down detailed rules for the implementation of European Parliament and Council Regulation (EC) No 178/2002 with regard to the network of organisations operating in the fields within the European Food Safety Authority's mission. Off J Eur Union L379:64–67

European Commission (2006) Commission Regulation (EC) No 575/2006 of 7 April 2006 amending Regulation (EC) No 178/2002 of the European Parliament and of the Council as regards the number and names of the permanent Scientific Panels of the European Food Safety Authority. Off J Eur Union L100:3–3

European Commission (2008) Commission Regulation (EC) No 202/2008 of 4 March 2008 amending Regulation (EC) No 178/2002 of the European Parliament and of the Council as regards the number and names of the Scientific Panels of the European Food Safety Authority. Off J Eur Union L60:17–17

European Commission (2011) Commission Implementing Regulation (EU) No 931/2011 of 19 September 2011 on the traceability requirements set by Regulation (EC) No 178/2002 of the European Parliament and of the Council for food of animal origin. Off J Eur Union L242:2–3

European Commission (2013) Commission Implementing Regulation (EU) No 208/2013 of 11 March 2013 on traceability requirements for sprouts and seeds intended for the production of sprouts. Off J Eur Union L68:16–18

European Parliament and Council (2002) Regulation (EC) No 178/2002 of the European Parliament and of the Council of 28 January 2002 laying down the general principles and requirements of food law, establishing the European Food Safety Authority and laying down procedures in matters of food safety. Off J Eur Comm L 31:1–24

European Parliament and Council (2003a) Regulation (EC) No 1642/2003 of the European Parliament and of the Council of 22 July 2003 amending Regulation (EC) No 178/2002 laying down the general principles and requirements of food law, establishing the European Food Safety Authority and laying down procedures in matters of food safety. Off J Eur Union L245:4–6

European Parliament and Council (2003b) Regulation (EC) No 1830/2003 of the European Parliament and of the Council of 22 September 2003 concerning the traceability and labelling of genetically modified organisms and the traceability of food and feed products produced from genetically modified organisms and amending Directive 2001/18/EC. Off J Eur Union L268:24–28

European Parliament and Council (2004) Regulation (EC) No 1935/2004 of the European Parliament and of the Council of 27 October 2004 on materials and articles intended to come into contact with food and repealing Directives 80/590/EEC and 89/109/EEC. Off J Eur Union L338:4–17

European Parliament and Council (2009) Regulation (EC) No 596/2009 of the European Parliament a n of the Council of 18 June 2009 adapting a number of instruments subject to the procedure referred to in Article 251 of the Treaty to Council Decision 1999/468/EC with regard to the regulatory procedure with scrutiny adaptation to the regulatory procedure with scrutiny—Part Four. Off J Eur Union L188:14–92

European Parliament and Council (2014) Regulation (EU) No 652/2014 of the European Parliament and of the Council of 15 May 2014 laying down provisions for the management of expenditure relating to the food chain, animal health and animal welfare, and relating to plant health and plant reproductive material, amending Council Directives 98/56/EC, 2000/29/EC and 2008/90/EC, Regulations (EC) No 178/2002, (EC) No 882/2004 and (EC) No 396/2005 of the European Parliament and of the Council, Directive 2009/128/EC of the European Parliament and of the Council and Regulation (EC) No 1107/2009 of the European Parliament and of the Council and repealing Council Decisions 66/399/EEC, 76/894/EEC and 2009/470/EC. Off J Eur Union L189: 1–32

FAO (2017) Food Traceability Guidance. Food and Agriculture Organization of the United Nations (FAO), Rome. Available http://www.fao.org/3/a-i7665e.pdf. Accessed 29 Sept 2020

Fiorino M, Barone C, Barone M, Mason M, Bhagat A (2019) The intentional adulteration in foods and quality management systems: chemical aspects. Quality systems in the food industry. Springer International Publishing, Cham, pp 29–37

FMRIC (2008) Handbook for introduction of food traceability systems (Guidelines for Food Trace-ability), March 2007, 2nd edn. Food Marketing Research and Information Center (FMRIC), Tokyo. Available http://www.maff.go.jp/j/syouan/seisaku/trace/pdf/handbook_en.pdf. Accessed 29 Sept 2020

Future of Fish, FishWise, and Global Food Traceability Center (2018) Seafood traceability glossary—a guide to terms, technologies, and topics. Future of Fish, FishWise, and Global Food Traceability Center. Available http://futureoffish.org/content/seafood-traceability-glossary-guide-terms-technologies-and-topics. 29 September 2020

Giraud G, Halawany R (2006) Consumers' perception of food traceability in Europe. Proceedings of the 98th EAAE Seminar, Marketing Dynamics within the Global Trading System, Chania, Greece

Golan E, Krissoff B, Kuchler F (2004a) Food traceability one ingredient in safe efficient food supply. Amber Waves 2(2):14–21

Golan EH, Krissoff B, Kuchler F, Calvin L, Nelson K, Price G (2004b) Traceability in the US food supply: economic theory and industry studies (No. 33939). Economic Research Service, United States Department of Agriculture, Agricultural Economic Report No. 830. United States Department of Agriculture, Washington, DC

GSFA (2017) Codex General Standard of Food Additives (GSFA) Online Database. Codex Alimentarius Commission, Rome. Available http://www.fao.org/gsfaonline/index.html;jsessionid=B20 A337D098220B1C4BB75A4B8AB3254. Accessed 29 Sept 2020

Haddad MA, Dmour H, Al-Khazaleh JFM, Obeidat M, Al-Abbadi A, Al-Shadaideh AN, Al-mazra'awi MS, Shatnawi MA, Iommi C (2020a) Herbs and medicinal plants in Jordan. J AOAC Int 103(4):925–929. https://doi.org/10.1093/jaocint/qsz026

Haddad MA, El-Qudah J, Abu-Romman S, Obeidat M, Iommi C, Jaradat DSM (2020b) Phenolics in Mediterranean and Middle East important fruits. J AOAC Int 103(4):930–934. https://doi.org/10.1093/jaocint/qsz027

Haddad MA, Parisi S (2020a) Evolutive profiles of Mozzarella and vegan cheese during shelf-life. Dairy Ind Int 85(3):36–38

Haddad MA, Parisi S (2020b) The next big HITS. New Food Magazine 23(2):4

Herzallah SM (2012) Detection of genetically modified material in feed and foodstuffs containing soy and maize in Jordan. J Food Comp Anal 26(1–2):169–172. https://doi.org/10.1016/j.jfca.2012.01.007

Hoorfar J, Jordan K, Butler F, Prugger R (eds) (2011) Food chain integrity: a holistic approach to food traceability, safety, quality and authenticity. Woodhead Publishing, Cambridge, Philadelphia, and New Delhi

Hosch G, Blaha F (2017) Seafood traceability for fisheries compliance—country-level support for catch documentation schemes. FAO Fisheries and Aquaculture Technical Paper 619. Food and Agriculture Organization of the United Nations (FAO), Rome, pp 1–114. Available http://www.fao.org/3/a-i8183e.pdf. Accessed 29 Sept 2020

IFT (2018) Global Food Traceability Center, Glossary of Terms. Institute of Food Technologists (IFT), Chicago. Available http://www.ift.org/gftc/glossary-of-terms.aspx. Accessed 29 Sept 2020

Italian Institute of Packaging (2011) Aspetti analitici a dimostrazione della conformità del food packaging: linee guida. Prove, calcoli, modellazione e altre argomentazioni. Istituto Italiano Imballaggio, Milan

Jasnen-Vullers MH, van Dorp CA, Beulens AJM (2003) Managing traceability information in manufacture. Int J Inf Manag 23(5):395–413. https://doi.org/10.1016/S0268-4012(03)00066-5

Jin S, Zhang Y, Xu Y (2017) Amount of information and the willingness of consumers to pay for food traceability in China. Food Control 77:163–170. https://doi.org/10.1016/j.foodcont.2017.02.012

Kawecka A (2014) BRC/IOP standard importance in packaging quality assurance. Prod Eng Arch 4(3):14–17. https://doi.org/10.30657/pea.2014.04.04

Kim M, Hilton B, Burks Z, Reyes J (2018) Integrating blockchain, smart contract-tokens, and IoT to design a food traceability solution. In: Proceedings of the 2018 IEEE 9th Annual Information Technology, Electronics and Mobile Communication Conference (IEMCON), pp 335–340

King T, Cole M, Farber JM, Eisenbrand G, Zabaras D, Fox EM, Hill JP (2017) Food safety for food security: relationship between global megatrends and developments in food safety. Trend Food Sci Technol 68:160–175. https://doi.org/10.1016/j.tifs.2017.08.014

Kok E (2017) DECATHLON–development of cost efficient advanced DNA-based methods for specific traceability issues and high level on-site application–FP7 project. Impact 3:42–44. https://doi.org/10.21820/23987073.2017.3.42

Lacorn M, Lindeke S, Siebeneicher S, Weiss T (2018) Commercial ELISA measurement of Allergens and Gluten: what we can learn from case studies. J AOAC Int 101(1):102–107. https://doi.org/10.5740/jaoacint.17-0399

Lauge A, Sarriegi J, Torres J (2008) The dynamics of crisis lifecycle for emergency management. Proceedings of the 27th International Conference of the System Dynamic Society, 2009. System Dynamic Society, Albuquerque

Lewis KE, Grebitus C, Colson G, Hu W (2016) German and British consumer willingness to pay for beef labeled with food safety attributes. J Agric Econ 68(2):451–470. https://doi.org/10.1111/1477-9552.12187

Mania I, Barone C, Caruso G, Delgado A, Micali M, Parisi S (2016a) Traceability in the cheese-making field. the regulatory ambit and practical solutions. Food Qual Mag 3:18–20. ISSN 2336-4602

Mania I, Fiorino M, Barone C, Barone M, Parisi S (2016b) Traceability of packaging materials in the cheesemaking field. the EU Regulatory Ambit. Food Packag Bull 25(4&5):11–16

Mania I, Delgado AM, Barone C, Parisi S (2018) Traceability in the dairy industry in Europe. Springer International Publishing, Heidelberg, Germany

Meuwissen MP, Velthuis AG, Hogeveen H, Huirne R (2003) Traceability and certification in meat supply chains. J Agribus 21(2):167–181

Mitroff II, Shrivastava P, Udwadia FE (1987) Effective crisis management. Acad Manag Execut 1(3):283–292

Moyer DC, DeVries JW, Spink J (2017) The economics of a food fraud incident–case studies and examples including melamine in wheat gluten. Food Control 71:358–364. https://doi.org/10.1016/j.foodcont.2016.07.015

Nicolae CG, Moga LM, Bahaciu GV, Marin MP (2017) Traceability system structure design for fish and fish products based on supply chain actors needs. Scientific Papers: Series D, Animal Science 60:353–358. ISSN 2285-5750

Nykamp DQ (2018) Node degree definition. Math Insight. Available http://mathinsight.org/definition/node_degree. Accessed 29 Sept 2020

Olsen P, Borit M (2018) The components of a food traceability system. Trend Food Sci Technol 29(2):142–150. https://doi.org/10.1016/j.tifs.2012.10.003

Parisi S (2002) I fondamenti del calcolo della data di scadenza degli alimenti: principi ed applicazioni. Ind Aliment 41(417):905–919

Parisi S (2009) Intelligent packaging for the food industry. In: Carter EJ (ed) Polymer electronics—a flexible technology. Smithers Rapra Technology Ltd, Shawbury

Parisi S (2012a) Food industry and food alterations. the user-oriented approach. Smithers Rapra Technologies, Shawsbury

Parisi S (2012b) La mutua ripartizione tra lipidi e caseine nei formaggi. Un approccio simulato. Ind Aliment 523(51):7–15

Parisi S (2013) Food Industry and packaging materials—performance-oriented guidelines for users. Smithers Rapra Technologies, Shawsbury

Parisi S (2016) The world of foods and beverages today: globalization, crisis management and future perspectives. Learning.ly/ The Economist Group. Available http://learning.ly/products/the-world-of-foods-and-beverages-today-globalization-crisis-management-and-future-perspectives

Parisi S (2019) Analysis of major phenolic compounds in foods and their health effects. J AOAC Int 102(5):1354–1355. https://doi.org/10.5740/jaoacint.19-0127

Parisi S (2020) Characterization of major phenolic compounds in selected foods by the technological and health promotion viewpoints. J AOAC Int, in press. https://doi.org/10.1093/jaoacint/qsaa011

Parisi S, Barone C, Sharma RK (2016) Chemistry and food safety in the EU. The rapid alert system for food and feed (RASFF). Springer briefs in molecular science: Chemistry of foods, Springer

Perreten V, Schwarz F, Cresta L, Boeglin M, Dasen G, Teuber M (1997) Nature 389(6653):801–802. https://doi.org/10.1038/39767

Phillips I (2003) Does the use of antibiotics in food animals pose a risk to human health? A critical review of published data. J Antimicrob Chemother 53(1):28–52. https://doi.org/10.1093/jac/dkg483

Piergiovanni L, Limbo S (2016) Food packaging materials. Springer International Publishing, Cham. https://doi.org/10.1007/978-3-319-24732-8

Pisanello D (2014) Chemistry of foods: EU legal and regulatory approaches. Springer briefs in chemistry of foods. Springer International Publishing, Cham

Pisanello D, Caruso G (2018) Novel foods in the European Union. Springer International Publishing, Cham. https://doi.org/10.1007/978-3-319-93620-8

Schwägele F (2005) Traceability from a European perspective. Meat Sci 71(1):164–173. https://doi.org/10.1016/j.meatsci.2005.03.002

Sheenan JJ (2007a) Acidification—19, What problems are caused by antibiotic residues in milk? In: McSweeney PLH (ed) Cheese problems solved. Woodhead Publishing Limited, Cambridge, and CRC Press LLC, Boca Raton

Sheenan JJ (2007b) Salt in cheese—46, How does NaCl affect the microbiology of cheese? In: McSweeney PLH (ed) Cheese problems solved. Woodhead Publishing Limited, Cambridge, and CRC Press LLC, Boca Raton

Silva SV, Malcata FX (2000) Action of cardosin a from Cynara humilis on ovine and caprine caseinates. J Dairy Res 67(3):449–454. https://doi.org/10.1017/s0022029900004234

Silvis ICJ, van Ruth SM, van der Fels-Klerx HJ, Luning PA (2017) Assessment of food fraud vulnerability in the spices chain: an explorative study. Food Control 81:80–87. https://doi.org/10.1016/j.foodcont.2017.05.019

Sodano V, Verneau F (2004) Traceability and food safety: public choice and private incentives, quality assurance, risk management and environmental control in agriculture and food supply networks. Proceedings of the 82nd Seminar of the European Association of Agricultural Economists (EAAE), Bonn, Germany, 14–16 May, Volumes A and B

Starbird SA, Amanor-Boadu V (2006) Do inspection and traceability provide incentives for food safety? J Agric Res Econ 31(1):14–26

Steinka I, Parisi S (2006) The influence of cottage cheese manufacturing technology and packing method on the behaviour of micro- flora. Joint Proc, Deutscher Speditions- und Logistikverband e.V., Bonn, and Institut für Logistikrecht & Riskmanagement, Bremerhaven

Stilo A, Parisi S, Delia S, Anastasi F, Bruno G, Laganà P (2009) La Sicurezza Alimentare in Europa: confronto tra il 'Pacchetto Igiene' e gli Standard British Retail Consortium (BRC) ed International Food Standard (IFS). Ann Ig 21(4):387–401

Telesetsky A (2017) US seafood traceability as food law and the future of marine fisheries. Environ Law 43(3):765–795

Vairo Cavalli S, Silva SV, Cimino C, Malcata FX, Priolo N (2008) Hydrolysis of caprine and ovine milk proteins, brought about by aspartic peptidases from Silybum marianum flowers. Food Chem 106(3):997–1003. https://doi.org/10.1016/j.foodchem.2007.07.015

Verbeke W, Frewer LJ, Scholderer J, De Brabander HF (2007) Why consumers behave as they do with respect to food safety and risk information. Anal Chim Acta 586(1–2):2–7. https://doi.org/10.1016/j.aca.2006.07.065

Zanoli R, Naspetti S (2002) Consumer motivations in the purchase of organic food. a means-end approach. British Food J 104(8):643–653

Zhang D (2015) Best practices in food traceability. In: Proceeding ot the Institute of Food Technologists (IFT) 15, Chicago, 10 July–13 July 2015. Available http://www.ift.org/gftc/~/media/GFTC/Events/Best%20Practices%20in%20Food%20Traceability.pdf. Accessed 29 Sept 2020

Chapter 2
Traditional Foods in Jordan and Traceability. *Hummus* and Related Variations

Abstract In the Middle East area, there are different products which could be discussed when speaking of authenticity and traceability, above all with reference to the difference between 'hand-made' or 'artisanal' food as opposed to 'industrial' or 'made with non-traditional ingredients'. In addition, traceable food products may be also an interesting legacy for many geographical and ethnic cultures. Consequently, the examination of certain recipes or food products linked with history and Mediterranean traditions can be really interesting when speaking of food traceability. In this ambit, Jordanian foods can represent a peculiar case study. Some of these traditional products are examined by different viewpoints, with peculiar attention to chemical composition, identification of raw materials, preparation procedures, and traceability. This chapter is dedicated to a peculiar Middle East dish which can be easily found in Jordanian markets: the *hummus* (and related versions). Because of the increasing success of *hummus* and *hummus*-like products worldwide, traceability alternatives should be analysed and discussed.

Keywords Authenticity · Food safety · *Hummus* · Jordan · Ethnic food · *Tahini* · Traceability

Abbreviations

Activity water	A_w
European Union	EU
Food and beverage	F&B
Food Business Operator	FBO
Lactic acid bacteria	LAB
Middle East	ME
Off-Line	OL
United States of America	USA

2.1 Food Traceability in Jordan. A General Overview

As mentioned in Sect. 1.1, the modern industry of foods and beverages (F&B) is obliged to solve problems concerning public hygiene, food safety, regulatory requirements, traceability and authenticity questions, etc. (Delgado et al. 2016a, b–2017; Mania et al. 2016a, b–2018; Parisi 2002, 2016; Parisi et al. 2016; Perreten et al. 1997; Phillips 2003; Pisanello 2014). In general, the list of most known risks and concerns is summarised as follows:

a. Microbiological risks in terms of food safety and commercial requirements;
b. Chemical risks affecting food safety;
c. Detection of foreign substances in foods (metals, plastics materials, wooden materials, nanochemicals, etc.);
d. Other food-related risks affecting consumers' health;
e. Demonstrable evidence of risk assessment (safety, integrity, and legal designation of food and beverage products);
f. Reliable evidence of continuous improvement (in terms of standard operative procedures, good manufacturing practices, etc.).

The problem of food authenticity—in other terms, the correspondence between the food description and its real nature—is extremely challenging at present, and it should be solved by means of many instruments, from analytical chemistry to software management of data (Al-Tal 2012; Food Standards Agency 2018; Johnson and Baumann 2010; Kvasnička 2005; Lunardo and Guerinet 2007). Anyway, two basic concepts have to be considered:

1. The match between description and content (of the F&B product) has to be assured, and
2. This concern can be only expressed (and consequently solved) only if the F&B product is really placed on the market.

The second concept highlights the role of 'unaware' consumers and aware F&B companies and food business operators (FBO). Food and beverage consumers can feel that traceability is a value (Buhr 2003; Cheek 2006; Chrysochou et al. 2009; Giraud and Halawany 2006; Golan et al. 2004; Jasnen-Vullers et al. 2003; Sodano and Verneau 2004; Starbird and Amanor-Boadu 2006; Verbeke et al. 2007). With reference to FBO, these stakeholders (Chapter 1) act as customers using F&B items for subsequent reworking or similar operations. In this ambit, traceability can be a powerful instrument (Epelbaum and Martinez 2014; Fiorino et al. 2019; Herzallah 2012; Hoorfar et al. 2011; Kim et al. 2018; Meuwissen et al. 2003; Schwägele 2005). Also, it has been considered that claimed authenticity is able to orient consumers' choices (also with subliminal messages), with undoubtable effects on the economy of local markets (Al-Tal 2012; DeSoucey 2010). Consequently, food traceability has to be managed correctly (Mania et al. 2016a, b–2018), depending on the 'active player' or 'subject' (Parisi 2016; Haddad and Parisi 2020a, b).

In the Middle East (ME) geographical area, there are different products which could be discussed when speaking of authenticity and traceability, above all with relation to the difference between 'hand-made' or 'artisanal' food as opposed to 'industrial' or 'made with non-traditional ingredients' product… This chapter is dedicated to a peculiar ME dish which can be easily found in Jordanian markets: the *hummus* (and related versions).

2.2   *Hummus*. General Features and History

The word *hummus* means generally a mixture of different ingredients, including dried chickpeas (*Cicer arietinum* L.), garlic, *tahini*,[1] lemon juice, and spices (El-Qudah 2015). It is widely diffused in the ME (Jordan, Palestine, Syria, Lebanon…), in Turkey, and also in non-Asian countries such as the United States of America (Yamani and Mehyar 2011). This recipe has been also linked with health effects in the context of the Mediterranean diet (Meneley 2007). Its use is reported alone and also as an appetiser, spread on flatbread such as the so-called Arabic *pita*.

From the social viewpoint, *hummus* is largely consumed in Jordan, as in many ME countries. The same situation can be observed also in particular groups such as Palestinian refugees settled in Jordan: according to a recent research, 49.2% of refugee people in Jordan eat *hummus* 1–2 times per week, while 11.2 and 25.6% are reported to consume this food daily or 3–4 times per week, respectively. For comparison purposes, it should be considered that *labaneh*—a product discussed in Chapter 3—is reported to be consumed at least 1–2 times per week by 88.2% of responding people (Tayyem 2010). This simple comparison demonstrates that *hummus* and also *labaneh* are really representative foods in the ME culture.

This situation and related preference are also partially dependent on nutritional values for this product. The list of ingredients contains boiled chickpea (approximately 20–25% of the total mass of raw materials), lemon juice (or citric acid, which is an important variable when speaking of traceability), garlic, salt, *tahini*, and spices (El-Qudah 2015; Yamani and Al-Dababseh 1994; Yamani and Mehyar 2011). Table 2.1 shows a simplified recipe from different sources.

From the historical viewpoint, *hummus* is reported to be a traditional product of Arab-Palestinian culture in the 1930s. At present, *hummus* products are known in other geographical areas (Raviv 2003). The importance of authentic *hummus* is also expressed by the name of specialised restaurants for *hummus* specialities only: *hummusia*. Also, Cyprus has its local version of *hummus*, named exactly *humoi* (χούμοι), as part of the local cuisine, and it is appreciated by both Turkish and Greek communities (Angastinioti and Hutchins-Wiese 2016). With relation to the United States of America (USA) and the European Union (EU), there are different *hummus* and *hummus*-like products used for different purposes (as a pizza topping or in form of chocolate food) (Kamila 2020).

---

[1] Also named *tahina* or *tahin*, it is a popular paste in ME, and also in Greece and East Asian cooking cultures. Ingredients are dehulled and roasted sesame seeds after milling, oil, and salt. The final *tahini* corresponds to an oily and viscous fluid mass (Isa 2001; Yamani and Mehyar 2011).

Table 2.1 Basic ingredients for traditional *hummus* and two alternative versions: the *hummus* without *tahini* and another product with lemon replaced with lime (*hummus*-like food), while normal chickpeas have been replaced with canned chickpeas (El-Qudah 2015; Yamani and Al-Dababseh 1994; Yamani and Mehyar 2011; Vegolosi.it 2020)

Traditional *Hummus*	*Hummus* without *tahini*	*Hummus*-like food
Chickpeas	Chickpeas	Canned chickpeas
Tahini (from: dehulled and roasted sesame seeds (after milling); oil; and salt	Diced dried tomatoes	*Tahini* (from: dehulled and roasted sesame seeds (after milling); oil; and salt
Lemon juice (or citric acid)	Oregano (replacing the *tahini* paste)	Lime (replacing lemon juice)
Salt	Lemon juice (or citric acid)	Salt
Garlic	Salt	Garlic

2.3 *Hummus*. Basic Raw Materials, Preparation, Chemical Features, and Microbial Profiles

The basic ingredient of *hummus* is chickpea. In fact, the Arabic word for chickpea is '*hummus*' (a variation of Arabic ḥimmaṣ or ḥimmiṣ (American Heritage Dictionary 2020). The internationally recognised *hummus* dish is a dish of pureed chickpeas with *tahini* (a paste prepared with sesame seeds) addition (Isa 2001). Historically, the first culinary recipe similar to *hummus* with *tahini* seems to be the *hummus kasā* (thirteenth century), showing the use of many ingredients: chickpeas, oil, *tahini*, pepper, mint, walnuts, pistachios, and other raw materials (Rodinson et al. 2001; Shaheen et al. 2019).

With relation to the basic *hummus* preparation, the following steps could be considered (Faris and Takruri 2002; Shaheen et al. 2019; Vegolosi.it 2020; Yamani and Mehyar 2011):

a. Soaking of chickpeas in water (for overnight or for 6–12 h);
b. Boiling of the intermediate mass in aqueous solution containing sodium bicarbonate until a soft texture (acceptable from the traditional viewpoint) is obtained. Time should be approximately 45 min, although a 2-hour duration has been reported (Vegolosi.it 2020);
c. Addition of *tahini* and remaining ingredients and mixing until the final product is obtained. Suggested amounts: 200 grams of *tahini* per 1000 grams of chickpea; citric acid: 0.3–1.4%, if present in the recipe; 29 grams of salt per kg of chickpea.

On these bases, it has to be noted that the only heating procedure is boiling (step b). As a result, the durability of *hummus* products is quite limited (1–7 days under refrigerated conditions) because of the concomitant absence of chemical preservatives and additives able to modify chemical and microbiological profiles for shelf-life extension purposes (Yamani and Al-Dababseh 1994; Yamani and Mehyar 2011). In these conditions, it is obvious that traditional *hummus* cannot be sold with extended durability (and the limitation to local markets is normal).

In fact, the final *hummus* preparation exhibits the following chemical features (Singh et al. 2004; Yamani and Al-Dababseh 1994; Yamani and Mehyar 2011):

1. pH value: approximately 5.1;
2. Water activity (a_w) > 0.98.

In addition, the bioavailable concentration of carbohydrates (oligosaccharides) is really notable. In these conditions, the prevailing microbial flora is represented by lactic acid bacteria (LAB)—*Enterococcus, Leuconostoc,* and *Lactococcus* spp— while the presence of spreading microorganisms such as yeasts and several *Enterobacteriaceae* is reported to be normal (Yamani and Al- Dababseh 1994). Also, the contamination by safety menaces such as *E. coli, Salmonella,* and *L. monocytogenes* has to be taken into account (Shaheen et al. 2019). Should *hummus* products be found expired or spoiled, they show an aqueous texture with easily perceived sour aroma and some colorimetric changes (Shaheen et al. 2019; Yamani and Al-Dababseh 1994). Anyway, *hummus* has peculiar microbiological profiles because of its nature of microbiological culture medium.

Consequently, shelf life for traditional *hummus* is quite low and several attempts have been reported or suggested with the aim of extending durability: innovative metal or paper-based containers and chemical preservatives such as potassium sorbate, sodium metabisulfite, nisin, citric acid, thiamine dilauryl sulfate, and malic acid. In the case of added preservatives or microbial inhibiting agents, the aim is to limit the spread of yeasts in acidic foods such as *labaneh* (Chapter 3) and black olives (Al-Holy 2006; Amr and Yaseen 1994; Choi 2015; Shaheen et al. 2019; Mihyar et al. 1997, 1999; Turantaş et al. 1999; Scotter and Castle 2004; Jay et al. 2005; Henderson 2009).

As a consequence, the addition of chemical preservatives could affect the traditional feature of *hummus* products on the one side and naturally highlights a traceability problem because of the introduction of new ingredients, and each new raw material is potentially able to make more complex the procedure of traceability in each context. Another strategy has been suggested in the preparation of *hummus*: a sudden storage of softened and blended chickpea paste before mixing at 4 °C for about 10 h. This treatment has been reported to be useful for durability enhancement purposes (Shaheen et al. 2019).

Another problem could concern the use of canned chickpeas (because the first step, soaking, is not needed) (Vegolosi.it 2020). From the traceability viewpoint, it is obvious that canned chickpeas may simplify the traceability procedure. On the other hand, the origin of canned chickpeas has to be clarified.

The matter of *hummus* versions should also be considered. Three recipes (Vegolosi.it 2020) are shown here with the aim of demonstrating that *hummus* has spread worldwide (Edge 2010), although alternative versions may present some authenticity and traceability problems… As a result, a *hummus* without *tahini* may be suggested with the addition of diced dried tomatoes and oregano (these ingredients replace *tahini*). Another non-traditional version (*hummus*-like product, naturally) may be proposed with the complete replacement of chickpeas with young and green soybeans (*edamame*). The replacement of chickpeas may be also proposed (spicy *hummus*-like food) when speaking of the use of black beans, while lemon is replaced with lime (Pateman and El-Hamamsy 2003; Rombauer et al. 2002; Vegolosi.it 2020).

These alternatives (there are more than three single recipes, and the list could be long and not exhaustive!) justify the need to analyse the flow of input and output data with dedicated examples (Sects. 2.4 and 2.5). Also, the production of organic *hummus* needs to be demonstrated because of associated high prices is compared with non-organic claimed foods. Anyway, traceability examples shown here can take into account many possibilities. We have chosen four possibilities for example purposes.

From the nutritional viewpoint, and taking into account the variability of certain data based on the different recipes, it has been reported that commercial *hummus* provides approximately 166 calories (or 695 kJ) per 100 grams; the nutritional profile from the chemical angle should be summarised as follows (NutritionData 2020a):

• Carbohydrates: 14.3 grams;
• Dietary fibres: 6.0 g;
• Starch and sugars: unavailable data;
• Total fat: 9.6 grams (detail: saturated fat, 1.4 g; monounsaturated fat, 4.0 g; polyunsaturated fat, 3.6 g;
• Protein: 7.9 grams;
• Water: 66.6%.

In addition, vitamin B_6 (0.2 μg) and several minerals should be considered with attention.

The homemade *hummus* should have a similar profile (NutritionData 2020b):

• Energy intake: 177 calories (741 kJ);
• Carbohydrates: 20.1 grams;
• Dietary fibres: 4.0 g;
• Starch and sugars: unavailable data;
• Total fat: 8.6 grams (detail: saturated fat, 1.1 g; monounsaturated fat, 4.9 g; polyunsaturated fat, 2.1 g;
• Protein: 4.9 grams;
• Water: 64.9%.

In addition, vitamin B_6 (0.2 μg) and several minerals should be considered with attention.

Interestingly, the main difference between commercial and homemade preparations seems to be related to dry matter (100 grams of water): the higher the aqueous amount, the lower the amount of carbohydrates, and vice versa.

Naturally, these data should be considered with attention because of the possibility of alternative recipes. However, reported data show the notable abundance of carbohydrates, the most distinctive feature of *hummus* (and other chickpea-derived foods, in the ME and also in the Mediterranean basin) (Barone and Pellerito 2020; Bishouty 2000).

2.4 *Hummus*. The Flow of Input and Output Data

In the ambit of food traceability, the flow of input data should be based on a defined recipe. However, there are many possibilities when speaking of *hummus*. and Jordanian products may be only one of the many possibilities. Because of the importance of *hummus* in the ME, the Jordanian consumer should be aware of the origin of basic and minor raw ingredients when speaking of *hummus* preparations. Consequently, we have decided to discuss (Table 2.1) some hypothetical example based on three possible recipes, taking into account (Pateman and El-Hamamsy 2003; Rombauer et al. 2002):

a. The traditional recipe;
b. The alternative *hummus* without *tahini;*
c. The alternative and non-traditional version (*hummus* product) with lemon replaced with lime (*hummus*-like food), while normal chickpeas have been replaced with canned chickpeas (the first step, soaking, is not needed).

2.4.1 The Traditional Hummus

As explained in Sect. 2.3, the list of ingredients (Table 2.1) for traditional *hummus* (Fig. 2.1) should be summarised as follows:

a. Chickpeas for the preparation of puree;
b. *Tahini* (it has to be prepared apart). Ingredients for this paste are dehulled and roasted sesame seeds (after milling); oil (no peculiar preferences); and salt. The final *tahini* corresponds to an oily and viscous fluid mass;
c. Lemon juice (or citric acid);
d. Salt;
e. Garlic.

Fig. 2.1 Traditional *hummus* dish in Jordan

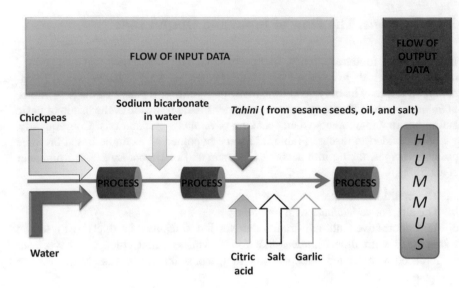

Fig. 2.2 Flow of input and output data for a traditional *hummus* recipe

In this ambit, the use of other ingredients is excluded. As a consequence, the flow of input data (Fig. 2.2) concerns the following materials:

1. The preparation of *tahini* (sesame seeds, oil, and salt) has to be considered as a pre-step before soaking (Sect. 2.3).
2. The soaking step considers only chickpeas.
3. The boiling step considers only softened chickpeas, water, and sodium bicarbonate.
4. The subsequent step concerns the mixing of all ingredients—intermediate pureed chickpeas, *tahini*, lemon (or citric acid), salt, and garlic.

As a consequence, and considering, for example, that each ingredient comes from one specific origin only (Fig. 2.2):

a. There is one possible identification concerning chickpeas when speaking of producer, country of origin, quantity, lot, and expiration date.
b. There is one possible identification for lemon (or citric acid) when speaking of producer, country of origin, quantity, lot, and expiration date. It has to be considered that lemon juice and citric acid may be used at the same time. As a result, we would have two entering raw materials (with two producers, two origins, two lots, … and so on).
c. There is one possible identification concerning salt when speaking of producer, country of origin, quantity, lot, and expiration date; the same thing can be affirmed when speaking of garlic.
d. Finally, the *tahini* implies the concomitant entering of three ingredients: sesame seeds, oil, and salt. Consequently, each of these ingredients means three separated elements in the flow of input data.

Figure 2.2 shows the situation. As easily explained, the 'traditional' recipe (Fig. 2.2) may imply that a minimum of five origins are present, with a maximum number of eight country identifications. In other words, each ingredient may be different from other ingredients with relation to producer identification, origin, lot, etc., even if one of these raw materials is mentioned more than one time. As an example, salt is mentioned two times. However, there are two possibilities:

1. Salt is used both for *tahini* and the final *hummus*.
2. Alternatively, there are two different salts.

Moreover, it could be observed that *tahini* is prepared apart by an external producer. Consequently, the *hummus* producer may also take into account only: chickpeas; lemon and/or citric acid; salt; garlic; and *tahini*, without further specifications. The last possibility means only five ingredients and five consequent identifications, while the 'worst' option may consider eight possible ingredients. The difference is that the producer has to take into account '*n*' ingredients. Consequently, the traceability exercise should consider only raw materials as used by the producer without the need to investigate the previous history. Naturally, official investigations have to take into account the complete food chain, but this example concerns only the basic operations by one producer only.

2.4.2 The Alternative Hummus Without Tahini

The list of ingredients for the alternative traditional *hummus* without *tahini* (Table 2.1) is summarised as follows (Vegolosi.it 2020):

a. Chickpeas for the preparation of puree;
b. Diced dried tomatoes;
c. Oregano (replacing the *tahini* paste);
d. Lemon juice (or citric acid);
e. Salt.

In this ambit, the use of *tahini* and garlic is excluded, while diced dried tomatoes are included. As a consequence, the flow of input data (Fig. 2.3) concerns the following materials:

1. The soaking step considers only chickpeas and water.
2. The boiling step considers only softened chickpeas, water, and sodium bicarbonate.
3. The subsequent step concerns the mixing of all ingredients—intermediate pureed chickpeas, diced dried tomatoes, lemon (or citric acid), oregano, and salt.

As a consequence, and considering, for example, that each ingredient comes from one specific origin only (Fig. 2.3):

1. There is one possible identification concerning chickpeas when speaking of producer, country of origin, quantity, lot, and expiration date.

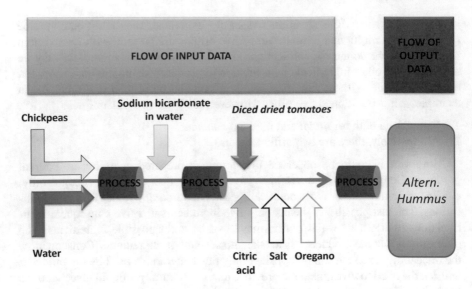

Fig. 2.3 Flow of input and output data for *hummus* without *tahini*

2. There is one possible identification for lemon (or citric acid) when speaking of producer, country of origin, quantity, lot, and expiration date. It has to be considered that lemon juice and citric acid may be used at the same time. As a result, we would have two entering raw materials (with two producers, two origins, two lots, … and so on).

3. There is one possible identification concerning salt when speaking of producer, country of origin, quantity, lot, and expiration date. The same thing can be affirmed when speaking of oregano and diced dried tomatoes.

Figure 2.3 shows the situation. As easily explained, this recipe may imply that a minimum of five origins are present, with a maximum number of six country identifications. In this situation, each raw material is mentioned only one time. However, the possible use of lemon juice and citric acid at the same time may complicate the flow of input data.

Interestingly, this flow appears simpler than the traditional situation. It has to be noted, on the other hand, that the presence of diced dried tomatoes is not on the same 'track' of traditional *hummus*, meaning that some 'foreign' ingredient entered the recipe. With relation to traceability and origins, there is the possibility that dried tomatoes are not from the ME including some unexpected complications if the prepared *hummus* is defined as an alternative version of the same recipe and with the same name. With exclusive relation to dried tomatoes, their use is found in Jordanian heritage; however, it is now diminished due to the availability of fresh tomatoes and the recent globalisation.

2.4.3 The 'Lime' Hummus

The list of ingredients for the alternative and non-traditional version (Table 2.1) with lemon replaced with lime (*hummus*-like food), while normal chickpeas have been replaced with canned chickpeas should be summarised as follows (Vegolosi.it 2020):

1. Canned chickpeas (the first step, soaking, is not needed);
2. *Tahini* (it has to be prepared apart). Ingredients for this paste are dehulled and roasted sesame seeds (after milling); oil (no peculiar preferences); and salt. The final *tahini* corresponds to an oily and viscous fluid mass;
3. Lime (replacing lemon juice);
4. Salt;
5. Garlic.

In this ambit, the use of other ingredients is excluded. As a consequence, the flow of input data (Fig. 2.4) concerns the following materials:

a. The preparation of *tahini* (sesame seeds, oil, and salt) has to be considered as a pre-step before soaking (Sect. 2.3).
b. The boiling step considers only canned chickpeas, water, and sodium bicarbonate.
c. The subsequent step concerns the mixing of all ingredients—intermediate pureed chickpeas, *tahini*, lime, salt, and garlic.

As a consequence, and considering, for example, that each ingredient comes from one specific origin only (Fig. 2.4):

1. There is one possible identification concerning canned chickpeas when speaking of producer, country of origin, quantity, lot, and expiration date.

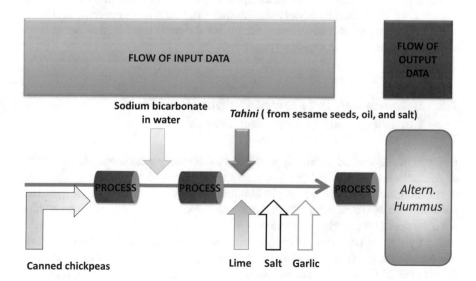

Fig. 2.4 Flow of input and output data for a *hummus*-like food with lime

2. There is one possible identification for lime when speaking of producer, country of origin, quantity, lot, and expiration date.
3. There is one possible identification concerning salt when speaking of producer, country of origin, quantity, lot, and expiration date; the same thing can be affirmed when speaking of garlic.
4. Finállly, the *tahini* implies the concomitant entering of three ingredients: sesame seeds, oil, and salt. Consequently, each of these ingredients means three separated elements in the flow of input data. The possibility of tahini adulteration with peanuts (because of low prices) has to be considered.

Figure 2.4 shows the situation. As easily explained, the 'non-traditional' recipe may imply that a minimum of five origins are present, with a maximum number of seven country identifications. In other words, each ingredient may be different from other ingredients with relation to producer identification, origin, lot, etc., even if one of these raw materials is mentioned more than one time. As an example, salt is mentioned two times. However, there are two possibilities:

a. Salt is used both for *tahini* and the final *hummus;*
b. Alternatively, there are two different salts.

Tahini may be also prepared apart by an external producer. Consequently, the *hummus* producer may also take into account only chickpeas; lime; salt; garlic; and *tahini*, without further specifications. The last possibility means only five ingredients and five consequent identifications, while the 'worst' option may consider seven possible ingredients. The difference with the traditional version is in the use of lime instead of lemon juice and/or citric acid. In addition, canned chickpeas have to be considered as an ingredient apart, but the identification of a similar ingredient may mean complications when speaking of country identifications. In our example, we have considered only one producer per each raw material. On a large-scale production, the use of different brands for the same ingredient is normal, and canned chickpeas are not surely an exception.

2.5 *Hummus* and the Flow of Output Data. Final Considerations Concerning Traceability

Because of the limited durability ascribed to *hummus* preparations, the durability of data concerning the final product (amount of produced *hummus*; lot, shelf life; destination; etc.) is necessarily brief. This situation may be an advantage for the whole food traceability chain because output data (correspondent to the final *hummus*) have not to be stored for notable period times. On the other hand, some industrial (sterilised into polycoupled packages or canned) *hummus* preparations are sold with shelf-life periods up to 60 days (4–7 days after opening, under refrigerated conditions) (Gregerson 2009). In other situations, the recommended refrigerated storage is 3–10 days (frozen product: 6–8 months) (Eat By Date 2020). The difference may be the presence of preservatives or new technological options, including packaging. Naturally, all these possibilities explain very well the need of a solid and robust food traceability chain of needed information, and this need should be particularly evident in the ME, while *hummus* is a traditional heritage and pride.

Additionally, the flow of output data in the upstream direction includes only the final product. Naturally, OL by-products should be declared (also because of their limited durability, with the possible not recommended use for other preparations unless a few hours only have passed and the OL materials have been stored under refrigerated conditions).

References

Al-holy M, Al-qadiri H, Lin M, Rasco B (2006) Inhibition of Listeria innocua in hummus by a combination of nisin and citric acid. J Food Prot 69(6):1322–1327. https://doi.org/10.4315/0362-028X-69.6.1322

Al-Tal SMS (2012) Modeling information asymmetry mitigation through food traceability systems using partial least squares. Electron J Appl Stat Anal 5(2):237–255

American Heritage Dictionary (2020) hum·mus also hum·us or hom·mos. The American Heritage Dictionnary of the English Language. Available https://ahdictionary.com. Accessed 1 Oct 2020

Amr AS, Yaseen EI (1994) Thermal processing requirements of canned chickpea dip. Int J Food Sci Technol 29:441–448. https://doi.org/10.1111/j.1365-2621.1994.tb02085.x

Angastinioti E, Hutchins-Wiese H (2016) Perception of and adherence to a mediterranean diet in cyprus and the United States. J Acad Nutr Diet 116(9):A88. https://doi.org/10.1016/j.jand.2016.06.318

Barone M, Pellerito A (2020) Palermo's street foods: The authentic Pane e Panelle. In: Sicilian Street Foods and Chemistry—The Palermo Case Study. Springer International Publishing, Cham. https://doi.org/10.1007/978-3-030-55736-2_5

Bishouty JI (2000) The nutritive value and safety of some selected street foods in Amman. Dissertation, University of Jordan, Amman

Buhr BL (2003) Traceability and information technology in the meat supply chain implications for firm organization and market structure. J Food Distrib Res 34(3):13–26. https://doi.org/10.22004/ag.econ.27057

Cheek P (2006) Factors impacting the acceptance of traceability in the food supply chain in the United States of America. Rev Sci Tech Off Int Epiz 25(1):313–319

Choi MR, Seul-Gi J, Qian L, Ga-Hee B, Su-Yeon L, Jeong-Woong P, Dong-Hyun K (2015) Effect of thiamine dilaurylsulfate against Escherichia coli O157:H7, Salmonella Typhimurium, Listeria monocytogenes and Bacillus cereus spores in custard cream. Food Sci Technol 60(1):320–324. https://doi.org/10.1016/j.lwt.2014.09.040

Chrysochou P, Chryssochoidis G, Kehagia O (2009) Traceability information carriers: The technology backgrounds and consumers' perceptions of the technological solutions. Appet 53(3):322–331. https://doi.org/10.1016/j.appet.2009.07.011

Delgado AM, Almeida MDV, Parisi S (2017) Chemistry of the mediterranean diet. Springer, Cham. https://doi.org/10.1007/978-3-319-29370-7

Delgado AM, Vaz de Almeida MD, Barone C, Parisi S (2016a) Leguminosas na Dieta Mediterrânica – Nutrição, Segurança, Sustentabilidade. CISA - VIII Conferência de Inovação e Segurança Alimentar, ESTM- IPLeiria, Peniche, Portugal

Delgado AM, Vaz de Almeida MD, Parisi S (2016b) Chemistry of the mediterranean diet. Springer, Cham. https://doi.org/10.1007/978-3-319-29370-7

DeSoucey M (2010) Gastronationalism: Food traditions and authenticity politics in the European Union. Am Sociol Rev 75(3):432–455. https://doi.org/10.1177/0003122410372226

Eat By Date (2020) How long does hummus last? https://www.eatbydate.com. Available https://www.eatbydate.com/other/snacks/hummus/. Accessed 29 Sept 2020

Edge JT (2010) Hummus catches on in America (as Long as It's Flavored). The New York Times, 15 June

El-Qudah JM (2015) Vitamin a contents per serving of eleven foods commonly consumed in Jordan. Int J Chem Tech Res 8(10):83–88

Epelbaum FMB, Martinez MG (2014) The technological evolution of food traceability systems and their impact on firm sustainable performance: A RBV approach. Int J Prod Econ 150:215–224. https://doi.org/10.1016/j.ijpe.2014.01.007

Faris MA, Takruri H (2002) Study of the effect of using different levels of tahinah (sesame butter) on the protein digestibility-corrected amino acid score (PDCAAS) of chickpea dip. J Sci Food Agric 83(1):7–12. https://doi.org/10.1002/jsfa.1273

Fiorino M, Barone C, Barone M, Mason M, Bhagat A (2019) The intentional adulteration in foods and quality management systems: chemical aspects. Quality systems in the food industry. Springer, Cham, pp 29–37

Food Standards Agency (2018) Packaging and labelling. Food Standards Agency, London. https://www.food.gov.uk. Available https://www.food.gov.uk/business-guidance/packaging-and-labelling#food-authenticity. Accessed 1 Oct 2020

Giraud G, Halawany R (2006) Consumers' perception of food traceability in Europe. In: Proceedings of the 98th EAAE Seminar, Marketing Dynamics within the Global Trading System, Chania, Greece

Golan E, Krissoff B, Kuchler F (2002) Traceability for food marketing and food safety: what's the next step? Agric Outlook 288:21–25

Golan E, Krissoff B, Kuchler F (2004) Food traceability one ingredient in safe efficient food supply. Amber Waves 2(2):14–21

Gregerson J (2009) How long after purchase will store-bought hummus remain good to eat? Shelf Life Advice LLC. Available http://shelflifeadvice.com/content/how-long-after-purchase-will-store-bought-hummus-remain-good-eat. Accessed 29 Sept 2020

Haddad MA, Parisi S (2020a) Evolutive profiles of Mozzarella and vegan cheese during shelf-life. Dairy Ind Int 85(3):36–38

Haddad MA, Parisi S (2020b) The next big HITS. New Food Magazine 23(2):4

Henderson P (2009) Sulfur dioxide: science behind this antimicrobial, anti- oxidant wine additive. Pract Winer 1:1–6. Available https://www.gencowinemakers.com/docs/Sulfur%20Dioxide-Science%20Behind%20this%20Anti-microbial,%20Anti-oxidant%20Wine%20Additive.pdf. Accessed 1 Oct 2020

Herzallah SM (2012) Detection of genetically modified material in feed and foodstuffs containing soy and maize in Jordan. J Food Comp Anal 26(1–2):169–172. https://doi.org/10.1016/j.jfca.2012.01.007

Hoorfar J, Jordan K, Butler F, Prugger R (eds) (2011) Food chain integrity: a holistic approach to food traceability, safety, quality and authenticity. Woodhead Publishing, Cambridge, Philadelphia, and New Delhi

Isa JK (2001) A study of the microbiological quality of tahina manufactured in Jordan. Dissertation, University of Jordan, Amman

Jasnen-Vullers MH, van Dorp CA, Beulens AJM (2003) Managing traceability information in manufacture. Int J Inf Manag 23(5):395–413. https://doi.org/10.1016/S0268-4012(03)00066-5

Jay JM, Loessner MJ, Golden DA (2005) Modern food microbiology, 7th Ed, Springer, New York, pp 301–442

Johnson J, Baumann S (2010) Foodies: democracy and distinction in the Gourmet. Foodscape, Routledge and New York

Kamila AY (2020) Hummus pizza is rising: a mash-up of middle eastern hummus with Italian pizza. Portland Press Herald. live. Available https://www.pressherald.com/2020/02/02/hummus-pizza-is-rising/. Accessed 29 Sept 2020

Kim M, Hilton B, Burks Z, Reyes J (2018) Integrating blockchain, smart contract-tokens, and IoT to design a food traceability solution. In: Proceedings of the 2018 IEEE 9th Annual Information Technology, Electronics and Mobile Communication Conference (IEMCON), pp 335–340

Kvasnička F (2005) Capillary electrophoresis in food authenticity. J Sep Sci 28(9–10):813–825. https://doi.org/10.1002/jssc.200500054

Lunardo R, Guerinet R (2007) The influence of label on wine consumption: its effects on young consumers' perception of authenticity and purchasing behavior. In: Proceedings of the 105th EAAE Seminar 'International Marketing and International Trade of Quality Food Products', Bologna, Italy, 8–10 March 2007

Mania I, Barone C, Caruso G, Delgado A, Micali M, Parisi S (2016a) Traceability in the Cheese-making Field. The Regulatory Ambit and Practical Solutions. Food Qual Mag 3:18–20. ISSN 2336-4602

Mania I, Delgado AM, Barone C, Parisi S (2018) Traceability in the dairy industry in Europe. Springer, Heidelberg, Germany

Mania I, Fiorino M, Barone C, Barone M, Parisi S (2016b) Traceability of packaging materials in the cheesemaking field. The EU regulatory ambit. Food Packag Bull 25, 4&5:11–16

Meneley A (2007) Like an extra virgin. Am Anthropol 109(4):678–687

Meneley A (2008) Oleo-signs and Quali-signs: the qualities of olive oil. Ethnos 73(3):303–326

Meuwissen MP, Velthuis AG, Hogeveen H, Huirne R (2003) Traceability and certification in meat supply chains. J Agribus 21(2):167–181

Mihyar GF, Yamani MI, Al-Sa'ed AK (1997) Resistance of yeast flora of labaneh to PS and SB. J Dairy Sci 80(10):2304–2309. https://doi.org/10.3168/jds.S0022-0302(97)76180-0

Mihyar GF, Yousif AK, Yamani MI (1999) Determination of benzoic and sorbic acids in labaneh by high-performance liquid chromatography. J Food Comp Anal 12(1):53–61. https://doi.org/10.1006/jfca.1998.0804

NutritionData (2020a) Hummus, commercial nutrition facts & calories. https://nutritiondata.self.com/. Available https://nutritiondata.self.com/facts/legumes-and-legume-products/4407/2. Accessed 29 Sept 2020

NutritionData (2020b) Hummus, homemade nutrition facts & calories. https://nutritiondata.self.com/. Available https://nutritiondata.self.com/facts/legumes-and-legume-products/4403/2. Accessed 29 Sept 2020

Parisi S (2002) I fondamenti del calcolo della data di scadenza degli alimenti: principi ed applicazioni. Ind Aliment 41(417):905–919

Parisi S (2016) The world of foods and beverages today: globalization, crisis management and future perspectives. Learning.ly/The Economist Group. Available http://learning.ly/products/the-world-of-foods-and-beverages-today-globalization-crisis-management-and-future-perspectives

Parisi S, Barone C, Sharma RK (2016) Chemistry and food safety in the EU.The rapid alert system for food and feed (RASFF). Springer briefs in molecular science: Chemistry of foods, Springer

Pateman R, El-Hamamsy S (2003) Egypt. Cavendish Square Publishing, Llc, New York

Perreten V, Schwarz F, Cresta L, Boeglin M, Dasen G, Teuber M (1997) Nature 389(6653):801–802. https://doi.org/10.1038/39767

Phillips I (2003) Does the use of antibiotics in food animals pose a risk to human health? A critical review of published data. J Antimicrob Chemother 53(1):28–52. https://doi.org/10.1093/jac/dkg483

Pisanello D (2014) Chemistry of foods: EU legal and regulatory approaches. Springerbriefs in chemistry of foods, Springer, Cham

Raviv Y (2003) Falafel: a national icon. Gastron 3(3):20–25. https://doi.org/10.1525/gfc.2003.3.3.20

Rodinson M, Arberry AJ, Perry C (2001) Medieval arab cookery. Prospect Books, Devon, p. 383

Rombauer IS, Rombauer Becker M, Becker E (2002) All about party foods and drinks. Charles Scribner's Sons, New York, p. 30

Schwägele F (2005) Traceability from a European perspective. Meat Sci 71(1):164–173. https://doi.org/10.1016/j.meatsci.2005.03.002

Scotter MJ, Castle L (2004) Chemical interactions between additives in food stuffs: a review. Food Addit Contam 21(2):93–124. https://doi.org/10.1080/02652030310001636912

Shaheen M, Nsaif M, Borjac Ja (2019) Effect of TDS on bacterial growth in Lebanese hummus dip. Beirut Arab Univ J Health Wellbeing 1, 2: Article 5. Available https://digitalcommons.bau.edu.lb/hwbjournal/vol1/iss2/5. Accessed 1 Oct 2020

Singh N, Sandhu KS, Kaur M (2004) Characterization of starches separated from Indian chickpea (Cicer arietinum L.) cultivars. J Food Eng 63(4):441–449. https://doi.org/10.1016/j.jfoodeng.2003.09.003

Sodano V, Verneau F (2004) Traceability and food safety: public choice and private incentives, quality assurance, risk management and environmental control in agriculture and food supply networks. In: Proceedings of the 82nd Seminar of the European Association of Agricultural Economists (EAAE), Bonn, Germany, 14–16 May, Volumes A and B

Starbird SA, Amanor-Boadu V (2006) Do inspection and traceability provide incentives for food safety? J Agric Res Econ 31(1):14–26

Tayyem RF (2010) Prevalence of food insecurity among palestinian refugees in Jordan. RHSC Working Paper Series, pp 1–25

Turantaş F, Göksungur Y, Dinçer A, Ünlütürk H, Güvenç A, Neşe Z (1999) Effect of potassium sorbate and sodium benzoate on microbial population and fermentation of black olive. J Sci Food Agric 79(9):1197–1202. https://doi.org/10.1002/(SICI)1097-0010(19990701)79:9%3C1197:AID-JSFA349%3E3.0.CO;2-A

Vegolosi.it (2020) Come fare l'hummus di ceci: ricetta e guida definitiva – VIDEO. Vegolosi.it. Available https://www.vegolosi.it/ricette-vegane/hummus/. Accessed 1 Oct 2020

Verbeke W, Frewer LJ, Scholderer J, De Brabander HF (2007) Why consumers behave as they do with respect to food safety and risk information. Anal Chim Acta 586(1–2):2–7. https://doi.org/10.1016/j.aca.2006.07.065

Yamani MI, Al-Dababseh BA (1994) Microbial quality of Hoummos (chickpea dip) commercially produced in Jordan. J Food Prot 57(5):431–435. https://doi.org/10.4315/0362-028X-57.5.431

Yamani MI, Mehyar GF (2011) Effect of chemical preservatives on the shelf life of hummus during different storage temperatures. Jordan J Agric Sci 7(1):19–31

Chapter 3
Jordan Dairy Products and Traceability. *Labaneh*, a Concentrated Strained Yogurt

Abstract The identification between food products and their claims is one of the most important challenges of the present time, with aspects ranging from marketing to export policy, from analytical chemistry to software management, and also with reference to food safety and public health. In addition, claimed authenticity is able to tacitly orient the choice of unaware consumers, with interesting effects on the economy of local markets, the identity of 'street foods' or 'ethnic foods', and the evolution of so-called 'Food Wars'. The Middle East area is an interesting laboratory when speaking of traditional and culturally linked foods and beverages. Middle East consumers are really interested in the origin, traceability, and authenticity of their traditional foods such as *hummus* and other foods with regional implications. This chapter is dedicated to a Middle East food which is common in Jordanian markets: *labaneh* (and related versions), with peculiar attention to chemical composition, identification of raw materials, preparation procedures, and traceability.

Keywords Authenticity · Ethnic food · In-bag straining · Jordan · *Labaneh* · Traceability · Yogurt

Abbreviations

Food and beverage	F&B
Generally recognised as safe	GRAS
Lactic acid bacteria	LAB
Middle East	ME
Off-Line	OL

3.1   Food Traceability in Jordan and in the Middle East. *Labaneh*

As mentioned in Sect. 2.1, the identification between food products and their claims is one of the most important challenges of the present time, with aspects ranging from marketing to export policy, from analytical chemistry to software management, and also with reference to food safety and public health (Al-Tal 2012; Food Standards Agency 2018; Johnson and Baumann 2010; Kvasnička 2005; Lunardo and Guerinet 2007). In addition, claimed authenticity is able to tacitly orient the choice of unaware consumers, with interesting effects on the economy of local markets, the identity of 'ethnic foods', and the evolution of so-called 'Food Wars' (Al-Tal 2012; Barone and Pellerito 2020; Bishouty 2000; DeSoucey 2010; Mania et al. 2016a, b, 2018; Parisi 2016–2019; Haddad and Parisi 2020a, b). Anyway, one of the key factors which can influence food and beverage consumers appears traceability, in a direct or indirect way (Buhr 2003; Cheek 2006; Chrysochou et al. 2009; Giraud and Halawany 2006; Golan et al. 2004; Jasnen-Vullers et al. 2003; Sodano and Verneau 2004; Starbird and Amanor-Boadu 2006; Verbeke et al. 2007).

The Middle East (ME) area is an interesting 'laboratory' when speaking of traditional and culturally linked foods and beverages. The situation of *hummus*—a food product very common in all ME countries, but also in Europe and in other continental areas such as the United States of America—should demonstrate that ME consumers are really interested in the origin, traceability, and authenticity (integrity) of their traditional foods (Chapter 2). The existence of regional variations in non-ME countries should be also considered. As a result, the higher the number of *hummus* variations, the higher the amount of traceability exercises (and possibilities). In addition, the higher the number of choices when speaking of raw materials, the higher the possible number of countries of origin with relation to one single ingredient (ME: Jordan, Lebanon, Palestine, …). The situation of *tahini* (Chapter 2) has been discussed. Moreover, the difference between 'hand-made' or 'artisanal' food as opposed to 'industrial' or 'made with non-traditional ingredients' products is always an important factor of the traceability matter. With reference to ME specialities, *hummus* is only one of possible foods with regional implications. This chapter is dedicated to a ME food which can be easily found in Jordanian markets: *labaneh* (and related versions).

3.2   *Labaneh*. General Features and History

The word *labaneh* means a peculiar ME dairy product similar to cow's milk yogurt. In fact, the semisolid food named *labaneh* represents set yogurt after the partial removal of acidic whey (Mihyar et al. 1999). *Labaneh*—also named *labneh* or *labenah*, depending on the peculiar country—derives from the Arabic name for milk: *laben*, representing of the most ME characteristic F&B expression of cultural heritage, also in the broad ambit of the Mediterranean diet (Delgado et al. 2017; El-Gendi 2015; Varnam and Sutherland 1994; Tamime and Robinson 1978). In fact, *labaneh* can be also found in Turkey, Greece (local name: 'greek yogurt'), the Balkans, Spain, India (local name: *shrikhand*), Iceland (local name: *skyr*), and North America (El-Salam et al. 2011; Rocha et al. 2014). Basically, this fermented milk product may be similar for some aspects to soft cottage, boursin, 'petit suisse', or quarg cheeses (Abu-Dieyed et al. 2007; Abu-Fisheh 1995; Al-Kadamany et al. 2002; Asad 2004; Hollingsworth 2001; Rocha et al. 2014; Tamime et al. 1989; Varnam and Sutherland 1994) because of the white/creamy aspect, the peculiar texture, sour taste, good spreadability, and a featured flavour depending on the presence of diacetyl (Al-Kadamany et al. 2002).

Nutritional profiles of this product should depend on the production method. The traditional *labaneh* (by means of the 'in-bag straining method' in cloth bags) should have a total solid content of 23–25%, while fat matter would be generally comprised between 8 and 11%. Interestingly, this fluid food is consumed in Jordan and in other ME countries with olive oils, or as a sandwich spread. These uses can be possible especially if *labaneh* has total solid contents below 25%, while higher solid contents (between 25 and 40%) may be a distinctive feature of unpackaged strained yogurts (these foods are dried up to 35–40%, rounded in a ball shape, and finally kept under olive oil in jars for six months or more), especially in areas with low urbanisation and without industrial plants (Salji 1991; Tamime and Robinson 1978).

Being this food a yogurt, it is expected that pH values are acidic (between 3.6 and 4.0) according to some study (Yamani and Abu-Jaber 1994; Tamime and Robinson 1978). As already discussed with reference to *hummus* (Chapter 2), the main problem of *labaneh* seems linked to durability. Packaged *labaneh* (into press-to-close plastic containers) may exhibit the possible spreading of spoilage microorganisms and psychrotrophic yeasts in particular, under refrigerated storage (Salji et al. 1987). In particular, it may be expected that packaged *labaneh* may exhibit a yeast count between 6.4 and 6.6 Log_{10} units (colony forming units per gram) (Mihyar et al. 1999; Yamani and Abu-Jaber 1994). Consequently, the use of food preservatives able to inhibit yeasts has been proposed. In detail, the use of sorbic and benzoic acids (also as sodium and potassium salts) has been proposed and cited because of their 'generally recognised as safe' (GRAS) status, according to the United States Food and Drug Administration (Boer and Nielsen 1995). In detail, traditional *labaneh* should have a shelf-life period of 14 days if stored at low temperature (such as 7 °C). For this reason, and because of the important (and probably unavoidable) contamination by yeasts such as *S. cerevisiae*, the pasteurisation of milk has been recommended, with the possible use of food additives.

With relation to pasteurisation, many cycles have been proposed so far: 82 °C for 16 s; 85–95° for 5–10–20 min, etc. Anyway, the problem of yeasts and the consequent durability (7–14–21 days under refrigerated storage) and the increasing marketing success of yogurts have progressively determined the suggestion of use of certain preservatives (sorbates and benzoates) with the aim of enhancing shelf-life values at least in industrial and large-scale produced foods. Another possible addition, suggested because of the success of low-fat foods, is the use of stabilisers (gelatine, guar gum, or xanthan gum) with the aim of making more easier the commercialisation of *labaneh* as a stable and possibly reduced-fat food.

Another important matter concerns the use of milk. Traditionally, *labaneh* is a yogurt obtained from cow's milk, but also from goat's or sheep's milk. The use of camel's milk has also been mentioned when speaking of *labaneh,* and the local economies of ME countries such as Jordan are in fact accustomed to use these milks in the production ambit of *labaneh* and also *jameed* (Carod Royo and Sánchez Paniagua 2015; Fuquay et al. 2011). With reference to traceability, the possibility of different milks has to be taken into account.

3.3   *Labaneh.* General Production Methods

The traditional 'in-bag straining' method for *labaneh* follows these steps (please note that several conditions may be varied depending on claimed traditional *labaneh* produced at a large-scale level (Yamani and Abu-Jaber 1994):

1. Heating treatment of (cow's, goat's, sheep's, etc.) milk at 95 °C for 5 min;
2. Cooling step (recommended temperature: 42–45 °C);
3. Addition of starter cultures such as 1–2% *Lactobacillus delbrueckii ssp. bulgaricus* and *Streptococcus salivarius sp. tbermophilus*. These starter cultures are generally used for yogurt production (Baglio 2014);
4. Incubation/microbial spreading at 42–45 °C until the production of acidic curd (duration: 3–4 h);
5. Second cooling;
6. Salt addition (0.9–1.0%) and kneading (Haddad et al. 2017);
7. Filling (in cloth bags);
8. Whey exclusion for gravity (duration: 8–48 h);
9. Final packaging in plastic containers.

This system has been repeatedly mentioned in the scientific literature. Naturally, it does not take into account the modern ('mechanical') *labaneh* production, based on centrifugation, ultrafiltration of normal yogurt, or the use of reverse osmosis treatments (Gharaibeh 2017; Ozer et al. 1998), which can be used with the aim of enhancing yields in protein with the consequent lowering of fat matter (Abu-Jdayil et al. 2002; Rocha et al. 2014). Anyway, being this system currently used—and claimed—when speaking of traditional *labaneh*, it can be used for our traceability exercises, taking into account some technological variation: the use of potassium sorbate and benzoate, or the addition of xanthan gum to the original recipe.

Anyway, the chemical profile of *labaneh* (in the traditional version) should approximately be compliant with the following data (Abu-Dieyed et al. 2007; Batshon 1980; Carod Royo and Sánchez Paniagua 2015; Fuquay et al. 2011):

• Total solids: 22–26%;
• Fat matter: 9–11%;

- Protein: 8.5–9.0%;
- Acidity value: 1.5–2.5%.

The problem with *labaneh* is related to the commercial typology. As above mentioned, two types can be placed on the market (Mohammed 2006; Tamime and Robinson 1978; Yamani and Abu-Jaber 1994):

a. A perishable *labaneh*, with total solids between 23 and 25%, and shelf life not exceeding 15 days at 6 °C or less, and
b. A stable *labaneh* (kept under olive oil), with total solids between 35 and 40%, and shelf life not exceeding nine months at 6 °C or less.

The existence of two dissimilar typologies has to be taken into account, when speaking of traditional *labaneh* and industrial typologies containing stabilisers and/or food preservatives (actually, the use of essential oils has been also proposed) (Al-Otaibi and El-Demerdash 2008; Ersöz et al. 2011; Ismail et al. 2006).

3.4 *Labaneh* and Traceability. The Flow of Input and Output Data

In the ambit of food traceability (Chapter 1), the flow of input and output data should be based on a defined recipe. Because of the importance of *labaneh* in the ME, it should be highlighted that the Jordanian consumer is generally aware of the origin of basic ingredients, while he/she could be unaware of minor ingredients added with the aim of extending durability and palatability/technological attributes of the product. As above explained, all unpleasant effects for *hummus* (and for *labaneh* also) may be shown as the spreading of certain microorganisms, yeasts above all. Consequently, the use of certain preservatives might be observed, although Jordanian Official Regulations do not allow their use (Gharaibeh 2017). On the other hand, the rheological properties and associated hedonistic features of *labaneh* (texture, viscosity, colour, creaminess, acid taste, peculiar flavour…) should be maintained during all the commercial life of the product (up to 15 days or nine months in Jordan, depending also on the amount of total solids, as stated in Sect. 3.2). Anyway, the basic flow of input and output data concerning traditional and industrial *labaneh* in Jordan should concern only a simple list of ingredients, without taking into account un-allowed ingredients, additives, preservatives, and also other food chemical compounds such as stabilisers which could affect the 'authenticity' of such a product. Consequently, we have decided to discuss the traditional recipe only.

3.4.1   The Traditional Labaneh

As explained in Sect. 3.2, the list of ingredients for traditional *labaneh* (Fig. 3.1) should be summarised as follows (Carod Royo and Sánchez Paniagua 2015; Fuquay et al. 2011; Tawalbeh et al. 2014; Yamani and Abu-Jaber 1994):

a. Crude (untreated) cow's milk (alternative possibilities: goat's, sheep's, camel's milk);

Fig. 3.1 Traditional *labaneh* in Jordan

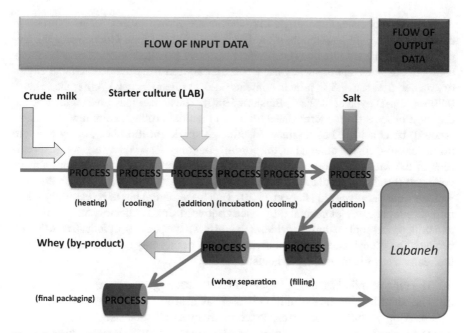

Fig. 3.2 Flow of input and output data for traditional *labaneh*

b. Selected lactic acid bacteria (LAB) as yogurt starter culture. Normally, the following LAB are considered in mixture (*L delbrueckii* ssp. *dulganbus* and *S. salivarius* sp. *thermophilus*);
c. Salt.

In this ambit, the use of other ingredients is excluded. As a consequence, the flow of input data (Fig. 3.2) concerns the following materials:

1. The preparation of milk concerns the following steps: initial heating (i.e. pasteurisation; first cooling; LAB inoculation; lactic fermentation); draining (expulsion of liquid whey in cloth bags under second cooling, overnight); and packaging in plastic pots;
2. The use of LAB has to be mandatorily done after the first cooling step. Interestingly, the lactic fermentation—absolutely critical—means that the intermediate fermented milk product cannot be altered by addition of other ingredients until the end of fermentation;
3. The addition of salt (recommended amount: 0.9–1.0%, although 2% may be observed and used usually in 35–40% moisture *labaneh* that will be stored up to 9 months under olive oils) has to be allowed after fermentation and before filling in cloth bags (generally, 20–25 kilograms should be filled in each bag). The operation may be done under second cooling; subsequently, the food product will be packaged in plastic pots and cooled (third cooling).

As a consequence, three single ingredients are shown in Fig. 3.2 as expression of input data, while output data should concern only the final *labaneh* and the by-product, also named Off-line (OL), represented by expelled whey. It may be interesting to consider that this whey has a peculiar composition by the chemical angle: dry matter, 5%; ash, 7.5%; protein content, 0.90%; sugar (lactose), 4.0%; lactic acid, 0.80% (Abu-Dieyed et al. 2007; Batshon 1980). These data may give an idea of the drainage process from intermediate yogurt to the final product. Each raw ingredient 'should' be considered as originated from one single location only, or by a single producer only. On the other hand, the Jordan economy—and many economical realities in the ME and in other areas—shows a situation while a notable amount of collected (crude) raw milk is from many small farmers, including households and women's cooperatives (Al Hiary et al. 2013). This situation has to be carefully taken into account because of general differences between small collectors on the one side and milk processors on the other side, especially when speaking of industrial *labaneh* productions. Consequently, the following critical points may be observed, in spite of the really short list of used ingredients:

a. The type of milk has to be considered (cow–sheep–goat–camel).
b. The identification of the milk collector, as always observed worldwide when speaking of milk cooperatives, concerns many small subjects.
c. The identification of the milk processor (*labaneh* producer) is not required generally when speaking of low-scale productions, at the farm level (Al Hiary et al. 2013). On the other side, a large-scale producer is responsible for the collection of physical raw milk and related identification data (by many milk collectors).
d. The identification of salt should be simple enough. It may be generally assumed that one producer only is required, and the general collection of data becomes easy. On the other hand, industrial plants may use different salt brands and types, with the consequent overload of traceability information.
e. Finally, the problem of starter culture has to be considered. In a 'large-scale production' ambit, it is generally observed that selected LAB cultures are from one single producer only, possibly from other countries, in Jordan and everywhere. Used LAB should be obtained only from the previous cultures regardless of the brand or company.

Figure 3.2 shows the situation. As easily explained, the 'traditional' recipe may imply a fixed number of ingredients (3), but the amount of collectors may be complex enough. Finally, the use of plastic containers—as part of the final product—has to be taken into account when speaking of traceability procedures. Official investigations have to take into account the complete food chain.

References

Abd El-Salam MH, Hippen AR, El-Shafie K, Assem FM, Abbas H, Abd El-Aziz M, Sharaf O, El-Aassar M (2011) Preparation and properties of probiotic concentrated yoghurt (labneh) fortified with conjugated linoleic acid. Int J Food Sci Technol 46(10):2103–2110. https://doi.org/10.1111/j.1365-2621.2011.02722.x

Abu-Dieyed ZHM, Al-Dabbas FM, Al-Dalain SYA (2007) Effect of drinking Labaneh whey on growth performance of broilers. Int J Poult Sci 6(11):842–845

Abu-Fisheh MA (1995) Concentration of acidic whey and its utilization in human bread making. Dissertation, University of Jordan

Abu-Jdayil B, Jumah RY, Shaker RR (2002) Rheological properties of a concentrated fermented product, labneh, produced from bovine milk: effect of production method. Int J Food Prop 5(3):667–679. https://doi.org/10.1081/JFP-120015500

Al Hiary M, Yigezu YA, Rischkowsky B, El-Dine Hilali M, Shdeifat B (2013) Enhancing the Dairy Processing Skills and Market Access of Rural Women in Jordan. International Center for Agricultural Research in the Dry Areas (ICARDA) Working Paper No. 565-2016-38922, pp. 1-12

Al-OtaibiM El-Demerdash H (2008) Improvement of the quality and shelf life of concentrated yoghurt (labneh) by the addition of some essential oils. Afr J Microbiol Res 2:156–161

Al-Kadamany E, Toufeili I, Khattar M, Abou-Jawdeh Y, Harakeh S, Haddad T (2002) Determination of shelf life of concentrated yogurt (Labneh) produced by in-bag straining of set yogurt using hazard analysis. J Dairy Sci 85(5):1023–1030. https://doi.org/10.3168/jds.S0022-0302(02)74162-3

Al-Tal SMS (2012) Modeling information asymmetry mitigation through food traceability systems using partial least squares. Electron J Appl Stat Anal 5(2):237–255

Angastinioti E, Hutchins-Wiese H (2016) Perception of and Adherence to a Mediterranean Diet In Cyprus and the United States. J Acad Nutr Diet 116(9):A88. https://doi.org/10.1016/j.jand.2016.06.318

Asad LA (2004) Investigation on heat-treatment and packaging of labaneh as a means of preservation. Dissertation, University of Jordan

Baglio E (2014) Chemistry and technology of yoghurt fermentation. SpringerBriefs in Chemistry of Foods, Springer, Cham, pp 25–33

Barone M, Pellerito A (2020) In: Sicilian street foods and Chemistry—the Palermo case study. Springer, Cham. https://doi.org/10.1007/978-3-030-55736-2

Batshon R (1980) Chemical and microbial analysis of Labaneh whey and its utilization for the production of single cell protein. Dissertation, University of Jordan, Amman

Bishouty JI (2000) The nutritive value and safety of some selected street foods in Amman. Dissertation, University of Jordan, Amman

Boer E, Nielsen PV (1995) Food preservatives. In: Samson RA, Hoekstra ES, Frisvad JC, Filtenborg O (eds) Introduction to food-borne fungi. Centraalbureau Voor Schimmelcultures, Utrecht

Buhr BL (2003) Traceability and information technology in the meat supply chain implications for firm organization and market structure. J Food Distrib Res 34(3):13–26. https://doi.org/10.22004/ag.econ.27057

Carod Royo M, Sánchez Paniagua L (2015) Estudio del efecto de aditivos en la calidad de un snack a base de labneh. Facultad de Veterinaria, Universidad Zaragoza. Available https://zaguan.unizar.es/record/37002/files/TAZ-TFG-2015-3990.pdf. Accessed 30 Sept 2020

Cheek P (2006) Factors impacting the acceptance of traceability in the food supply chain in the United States of America. Rev Sci Tech Off Int Epiz 25(1):313–319

Chrysochou P, Chryssochoidis G, Kehagia O (2009) Traceability information carriers. The technology backgrounds and consumers' perceptions of the technological solutions. Appet 53, 3:322–331. https://doi.org/10.1016/j.appet.2009.07.011

Delgado AM, Almeida MDV, Parisi S (2017) Chemistry of the mediterranean diet. Springer, Cham. http://dx.doi.org/10.1007/978-3-319-29370-7

DeSoucey M (2010) Gastronationalism: Food traditions and authenticity politics in the European Union. Am Sociol Rev 75(3):432–455. https://doi.org/10.1177/0003122410372226

El-Gendi MMN (2015) Comparative study between the microbiological quality of commercial and homemade labenah. Assiut Vet Med J 61, 147:148–153. Available http://www.aun.edu.eg/journal_files/437_J_547.pdf. Accessed 29 Sept 2020

Ersöz E, Kinik Ö, Yerlikaya O, Açu M (2011) Effect of phenolic compounds on characteristics of strained yoghurts produced from sheep milk. Afr J Agric Res 6(23):5351–5359. https://doi.org/10.5897/AJAR11.1012

Faris MA, Takruri H (2002) Study of the effect of using different levels of tahinah (sesame butter) on the protein digestibility-corrected amino acid score (PDCAAS) of chickpea dip. J Sci Food Agric 83(1):7–12. https://doi.org/10.1002/jsfa.1273

Food Standards Agency (2018) Packaging and labelling. Food Standards Agency, London, https://www.food.gov.uk. Available https://www.food.gov.uk/business-guidance/packaging-and-labelling#food-authenticity. Accessed 1 Oct 2020

Fuquay JW, Fox PF, McSweeney PLH (2011) Milk lipids. Encyclopedia of dairy sciences, vol 3, 2nd edn. Academic Press, Oxford, pp 649–740

Gharaibeh AA (2017) A Comparative Study of the Microbial, Physiochemical and Sensory Properties of Samples of Labneh Produced at Large (Industrial) Scale and Small Scale. Food Sci Qual Manag 63:1–6

Giraud G, Halawany R (2006) Consumers' Perception of Food Traceability in Europe. Proceedings of the 98th EAAE Seminar, Marketing Dynamics within the Global Trading System, Chania, Greece

Golan E, Krissoff B, Kuchler F (2002) Traceability for food marketing and food safety: what's the next step? Agric Outlook 288:21–25

Golan E, Krissoff B, Kuchler F (2004) Food Traceability One Ingredient in Safe Efficient Food Supply. Amber Waves 2(2):14–21

Haddad MA, Al-Qudah MM, Abu-Romman SM, Obeidat M, El-Qudah J, Al-Salt J (2017) Development of traditional jordanian low sodium dairy products. J Agric Sci 9(1):223–230. https://doi.org/10.5539/jas.v9n1p223

Haddad MA, Parisi S (2020a) Evolutive profiles of Mozzarella and vegan cheese during shelf-life. Dairy Industries International 85(3):36–38

Haddad MA, Parisi S (2020b) The next big HITS. New Food Magazine 23(2):4

Hassan HE, El Nemr TM (2007) Novel Recipes of Probiotic Low Fat Labneh Cheese as Natural Flavouring Salads in Hotel Meals. J Assoc Arab Univ Tour Hosp 4(1):131–146. https://doi.org/10.21608/jaauth.2007.68353

Henderson P (2009) Sulfur dioxide: science behind this antimicrobial, anti- oxidant wine additive. Pract Winer 1:1–6. Available https://www.gencowinemakers.com/docs/Sulfur%20Dioxide-Science%20Behind%20this%20Anti-microbial,%20Anti-oxidant%20Wine%20Additive.pdf. Accessed 1 Oct 2020

Hollingsworth P (2001) Yogurt reinvents itself. Food Technol 55:43–46

Ismail AM, Harby S, Salem AS (2006) Production of flavored labneh with extended shelf life. Egypt J Dairy Sci 34:59–68

Jasnen-Vullers MH, van Dorp CA, Beulens AJM (2003) Managing traceability information in manufacture. Int J Inf Manag 23(5):395–413. https://doi.org/10.1016/S0268-4012(03)00066-5

Johnson J, Baumann S (2010) Foodies: Democracy and Distinction in the Gourmet. Foodscape, Routledge, New York

Kvasnička F (2005) Capillary electrophoresis in food authenticity. J Sep Sci 28(9–10):813–825. https://doi.org/10.1002/jssc.200500054

Lunardo R, Guerinet R (2007) The influence of label on wine consumption: its effects on young consumers' perception of authenticity and purchasing behavior, Proceedings of the 105th EAAE Seminar 'International Marketing and International Trade of Quality Food Products', Bologna, Italy, 8–10 March 2007

Mania I, Barone C, Caruso G, Delgado A, Micali M, Parisi S (2016a) Traceability in the Cheese-making Field. The Regulatory Ambit and Practical Solutions. Food Qual Mag 3:18–20. ISSN 2336-4602

Mania I, Delgado AM, Barone C, Parisi S (2018) Traceability in the Dairy Industry in Europe. Springer International Publishing, Heidelberg, Germany

Mania I, Fiorino M, Barone C, Barone M, Parisi S (2016b) Traceability of Packaging Materials in the Cheesemaking Field. The EU Regulatory Ambit. Food Packag Bull 25, 4&5:11–16

Mihyar GF, Yamani MI, Al-Sa'ed AK (1997) Resistance of yeast flora of labaneh to PS and SB. J Dairy Sci 80(10):2304–2309. https://doi.org/10.3168/jds.S0022-0302(97)76180-0

Mihyar GF, Yousif AK, Yamani MI (1999) Determination of benzoic and sorbic acids in labaneh by high-performance liquid chromatography. J Food Comp Anal 12(1):53–61. https://doi.org/10.1006/jfca.1998.0804

Mohammed SYH (2006) The use of Lactococcus lactis in the production of labaneh. Dissertation, University of Jordan, Amman

Ozer BH, Bell AE, Grandison AS, Robinson RK (1998) Rheological Properties of Concentrated Yoghurt (Labneh). J Text Stud 29(1):67–79. https://doi.org/10.1111/j.1745-4603.1998.tb00154.x

Parisi S (2016) The world of foods and beverages today: globalization, crisis management and future perspectives. Learning.ly/The Economist Group. Available http://learning.ly/products/the-world-of-foods-and-beverages-today-globalization-crisis-management-and-future-perspectives

Parisi S (2019) Analysis of Major phenolic compounds in foods and their health effects. J AOAC Int 102(5):13541–355. https://doi.org/10.5740/jaoacint.19-0127

Ramos TM, Gajo AA, Pinto SM, Abreu LR, Pinheiro AC (2009) Perfil de textura de Labneh (iogurte Grego). Rev Inst Laticínios Cândido Tostes 64(369):8–12

Rocha DMUP, Martins JDFL, Santos TSS, Moreira AVB (2014) Labneh with probiotic properties produced from kefir: development and sensory evaluation. Food Sci Technol 34(4):694–700. https://doi.org/10.1590/1678-457x.6394

Salji JP (1991) Concentrated yoghurt: A challenge to our food industry. Food Sci Technol Today 5:18–19

Salji JP, Sawaya WN, Ayaz M (1987) The dairy processing industry in the central province of Saudi Arabia. Dairy Food Sanit 7:6–12

Sodano V, Verneau F (2004) Traceability and food safety: public choice and private incentives, quality assurance, risk management and environmental control in agriculture and food supply networks. Proceedings of the 82nd Seminar of the European Association of Agricultural Economists (EAAE), Bonn, Germany, 14–16 May, Volumes A and B

Starbird SA, Amanor-Boadu V (2006) Do Inspection and Traceability Provide Incentives for Food Safety? J Agric Res Econ 31(1):14–26

Tamime AY (1978) Concentrated yogurt 'Labneh'-a potential new dairy spread. Milk Ind 80(3):4–7

Tamime AY, Kalab M, Davies G (1989) Rheology and Microstructure of Strained Yoghurt (Labneh) Made From Cow's Milk by Three Different Methods. Food Str. 8(1):15

Tamime AY, Robinson RK (1978) Some aspects of the production of concentrated yogurt (Labaneh) popular in Middle East. Milk Sci Int 33:209–212

Tamime AY, Robinson RK (1999) Yoghurt science and technology, pp. 326–333. CRC Press, Boca Raton

Tawalbeh Y, Ajo R, Al-Udatt M, Gammoh S, Maghaydah S, Al-Qudah Y, Al-Sunnaq A, Al-Natour F (2014) Investigation of the antimicrobial preservatives in the dairy product (labneh). Food Sci Qual Manag 31:117–122

Varnam AH, Sutherland JP (1994) Milk and Milk Products. Chapman and Hall, London, Technology, Chemistry and Microbiology

Verbeke W, Frewer LJ, Scholderer J, De Brabander HF (2007) Why consumers behave as they do with respect to food safety and risk information. Anal Chim Acta 586(1–2):2–7. https://doi.org/10.1016/j.aca.2006.07.065

Yamani MI, Abu-Jaber MM (1994) Yeast flora of labaneh produced by in-bag straining of cow milk set yogurt. J Dairy Sci 77(12):3558–3564

Chapter 4
Dried Fermented Dairy Products in Jordan: *Jameed* and Traceability

Abstract The cultural heritage of several products should be re-discovered and studied with relation to many arguments, also including authenticity, food safety, and traceability. The Middle East area offers many traditional and culturally linked foods and beverages. Middle East consumers are really interested in the origin, traceability, and authenticity of their traditional foods such as *hummus, labaneh*, and other foods with regional implications. This Chapter deals with a peculiar Middle East food which can be easily found in Jordanian markets: *jameed*—also named *marees*—which is an alternative version of fermented food, very popular in Jordan and linked to the Jordanian national dish, *mansaf*. This spherical-like food, traditionally linked to the Bedouins, is a form of preserved and easily storable milk product, similarly to other non-milk based products present in other Mediterranean areas such as Sicily, and with a common history. This Chapter is dedicated to jameed, with peculiar attention to chemical composition, identification of raw materials, preparation procedures, and traceability.

Keywords Ethnic food · Freeze-drying · *Jameed* · Jordan · *Mansaf* · Solar-drying · Traceability

Abbreviations

LAB Lactic acid bacteria
ME Middle East
OL Off-Line

4.1 *Jameed*: A Solar-Dried Fermented Dairy Food

Jameed—also named *marees*—is an alternative version of ethnic food, very popular in the Middle East (Al Omari et al. 2008; Al-Qudah and Tawalbeh 2011). In detail, it is a fermented milk product widely diffused (and a cultural heritage) in Jordan and other Arabic Countries: Egypt (local name: *kishk*), Saudi Arabia, Syria, Iran, and Iraq (Al-Qudah and Tawalbeh 2011). This food is known in the Middle East (ME) and in many non-Arabic Countries because of its peculiarities as a fermented milk product, in form of 10–15 cm balls.

Jameed (Fig. 4.1) is traditionally linked to the Bedouins because its normal production period is the spring season, when milk collection is notable and some surplus can be obtained. Consequently, exceeding milk is preserved in form of dried fermented products. The spherical form has historical and practical reasons such as saving of storage spaces, easy packaging, and enhanced transportation options (Al-Qudah and Tawalbeh 2011). Interestingly, spherical shape is present today in other Mediterranean areas such as Sicily, when speaking of local *arancino* foods (rice balls). The origin of these 'street' and 'ethnic foods' is linked to the Arabic domination in Sicily (827–1,091 A.D.), and the spherical shape has been correlated with the need to have easily transportable foods (Barone and Pellerito 2020). With relation to *jameed*, the authenticity and traceability can be a key factor when speaking of

Fig. 4.1 Traditional *jameed* balls in Jordan

commercial success, similarly to other foods and beverages worldwide (Buhr 2003; Cheek 2006; Chrysochou et al. 2009; Giraud and Halawany 2006; Golan et al. 2004; Jasnen-Vullers et al. 2003; Sodano and Verneau 2004; Starbird and Amanor-Boadu 2006; Verbeke et al. 2007).

4.2   Production of *Jameed* Foods in the Middle East: Historical and Technological Reasons

The traditional and historical method for *jameed* production implies the use of (Al Omari et al. 2008):

a. Sheep's or goat's milk, although cow's and camel's milks have been reported
b. A bag made of goat hide (named *shikwa or sei'ein*), serving as a vat of milk collection and subsequent fermentation.

In accordance with the historical tradition, the fermentation process occurs naturally in the *shikwa* (or *sei'ein*) without addition of selected lactic acid bacteria (LAB) or other fermentation agents. In fact, the traditional procedure implies that the filled *shikwa* (or *sei'ein*) is used without preliminary washing. As a result, fermentation processes (duration: a few hours only) are the consequence of spreading of the active micro flora present in the bag. In fact, the Bedouins are accustomed to begin the daily production of fermented foods in the morning, and high environmental temperatures can guarantee the speed of fermentation processes on the one side, and the easy churning of resulting salted buttermilk (while butter is separated) on the other side. Subsequently, *jameed* is obtained by means of the mechanical (by hand) separation of whey from the fermented mass (Al Omari et al. 2008). The addition of salt is needed after this step. Subsequently, and differently from *labaneh*, the salted mass is shaped in form of pearl-like products by manual kneading: obtained single spherical units are covered with salt. *Jameed* balls have to be dried under shade for 24 h, and then placed under direct sunlight for 10–15 days with frequent salting and manual pressing to close the fractures and cracks in the balls, until the final result. This product has to show inner moisture < 20%, with 12% or more of salt. It should be noted that *jameed* balls are not hygroscopic, in spite of their protein and salt contents (Al Omari et al. 2008).

With relation to *jameed*, the simplest method recommends to strain buttermilk with adequate tools without preliminary heating, while the second one recommends the preliminary heating treatment (at 55–60° C, with the precipitation of protein concentrate at pH 4.6) of buttermilk before straining (Al Omari et al. 2008). In this ambit, the procedure is similar to the method used for *labaneh* Chapter 3, although the starting raw ingredient is milk (and not buttermilk) for the last mentioned product. After churning, the traditional method implies the drying procedure under direct sunlight; consequently, the name of 'solar-dried *jameed*' is justified also because of the meaning of Arabic *jamid* word: 'hard structure' (Al Omari et al. 2008).

Naturally, this system represents one of the aspects of the cultural heritage of the Bedouins. At present, there are other possibilities in Jordan, including small household productions and small-scale processes carried out by milk cooperatives when speaking of sheep's milk products (Iñiguez and Aw-Hassan 2005). As a consequence, chemical and microbiological profiles (Sect. 4.3), and also technological properties, may vary depending on the peculiar place and time of production, but also on technological features. In fact, drying performances depend on ball sizes (10–15 cm), drying times, and also on environmental conditions (temperature and exposure to sunlight), in the traditional method (Hilali et al. 2011).

On these bases, the final *jameed* has a low-moisture content (determined also by frequent 2%- salt addition) and a low pH value. Salt is added frequently for different drying time intervals, so the final *jameed* product will contain about 12% or more of salt. These conditions and the normal lactic fermentation giving lactic acid in the dried paste determine unfavourable conditions (in terms of positive redox potential and low water activity values) for the development of pathogens.

An alternative process for the production of *jameed* foods for experimental researches and purposes has been reported. Essentially, the design of the process is based on the 'solar-dried *jameed*' method, and basic differences concern (Al Omari et al. 2008):

a.   All steps are monitored when speaking of times, temperatures, and also the use of selected *Lactobacillus delbrueckii* culture media. The traditional culture addition is known as *laba*[1] (Al-Qudah and Tawalbeh 2011). Also, a stabilising additive such as carrageenan[2] may be used with the aim of improving texture and solubility in water (Quasem et al. 2009).

b.   The dimension of balls is defined (400 grams) with the aim of obtaining a constant drying in terms of physicochemical features, and also constant yields. Some research has studied the possible replacement of spherical shapes with other forms (Hamad et al. 2017). Moreover, *jameed* balls undergo a freeze-drying process.

c.   The solar light is replaced by natural convection solar dryers, and the drying time (15 days) is fixed until a constant weight is reached (Al Omari et al. 2008).

As a result, researchers found that processing alternative *jameed* products were a promising step if compared with traditional products, especially with reference to microbial profiles, commercial performances under storage, protection from environmental contamination (during the drying period under solar light), and the problem of oxidability because of observed superficial porosity of produced balls. In this ambit, it has been reported that porosity could be lower than expected depending on the use of caprine milk, because of higher casein aggregation and consequent enhanced

---

[1] *Laban* is a crude fermented yogurt containing *L. bulgaricus* and *S. thermophilus* as a mixed culture with other bacteria which is normally added by the Bedouins as the pure starter cultures of yogurt are not available for them.

[2] Carrageenan is used only for production of what is called liquid *jameed* in Jordan, excluding *jameed* balls. Liquid *jameed* is usually found in the markets and produced by dairy companies by liquefying *jameed* balls or making a concentrate of fermented skim milk.

hardness, although it has to be mentioned that this hypothesis is related to feta cheese from caprine milk (Hamad et al. 2016). Moreover, it should be mentioned that the higher yields of *jameed* from original milk appear to be obtained with the solar-dried process.

On the other hand, freeze-dried method is not so promising when speaking of good yields, if compared with the solar-drying procedure. With reference to good protein and fat amounts, the solar-dried system appears good enough.

In summary, it has been reported that the best nutritional and also antioxidant performances for *jameed* should be obtained by means of the solar-dried system with salt addition, while protein and fat profiles seem to augment if salt is not added (Alu'datt et al. 2016). Alternatively, 10–15% of whey protein powder may be added to *jameed* curd as recently reported (Ismail et al. 2017). In addition, it has been observed that different milks (sheep's or goat's milks are normally considered for the production of traditional *jameed*, although cow's or camel's milks may be used) may give different results in terms of total viable count. Interestingly, added LAB should contrast non-lactic and contaminant bacteria; however, it seems that TVC values are similar when speaking of fermentation without selected culture (Al Omari et al. 2008).

Other technologies such as spray-drying methods for *jameed* powders are reported (Hamad et al. 2016; Shaker et al. 2003), on condition that concentrated yogurt is partially defatted. In addition, different LAB such as *L. acidophilus*, *L. casei*, *L. kefir*, etc.—might be used as their application in other fermented foods is historically reported (Hamad et al. 2016; Ishnaiwer and Al-Razem 2013).

Anyway, the basic aims of these and other procedures are:

a. To ameliorate chemical composition of *jameed* products when speaking of non-constant protein and fat amounts (also, solubility should be improved in this way), and
b. To reduce contaminant bacteria in the final product.

With reference to the final use, it is obvious that *jameed*—differently from *labaneh*—is unsuitable as a spreadable product because of its notable hardness and low-moisture contents. On the contrary, it is normally crushed, then soaked in water, and subsequently used as a sauce (name: *sharap*) for traditional ME foods such as the national Jordanian dish, *mansaf* (Hamad et al. 2016).

From the technological and safety viewpoints, two features should be considered with attention, and taking into account that a general standard document concerning Jordanian *jameed* exists (JSI 1997):

a. The crushing and subsequent soaking in water determine a certain loss of nitrogen-based molecules (milk proteins) because of the difficult solubility of grinded *jameed* in water at 55° C. In fact, solubility is generally low if buttermilk is excessively heated (Quasem et al. 2009). In other terms, residues are normal after soaking, and they are not recovered. Consequently, an authentic *jameed* product should loss a certain amount of initial product when speaking of *jameed*-based sauces and this feature—unfavourable by the economic angle—should be taken into account when speaking of 'true' *jameed*

b. In addition, drying procedures are needed because *jameed* is a low-moisture product, and low water activity is surely a good feature because of the possible inhibition of dangerous microorganisms. On the other side, microbial contamination is possible during this critical step, and resident life forms may remain quiescent and living at the same time until the subsequent re-hydration process (water soaking), even after some months.

With relation to Jordanian standards, it has to be noted that moisture and sodium chloride should be low enough in the final *jameed* product (JSI 1997). In detail, moisture should be between 5 and 8% in well-dried *jameed* balls. These facts can also explain sensorial features of *jameed*: the taste is reported as salty and acidic, and long durability is generally estimated as one year (Hilali et al. 2011).

4.3 *Jameed:* Chemical and Microbiological Profiles

4.3.1 Chemical Profiles

From the nutritional viewpoint, *jameed* is classifiable as a high-protein food: in fact, protein content is generally between 48 and 54%, while fat matter is 7–10%. Because of low-moisture contents and the observed ash amount (12–13%), it has a naturally remarkable durability (Al Omari et al. 2008; Al-Saed et al. 2012; Mazahreh et al. 2008). From the viewpoint of traceability, *jameed* is linked to the type of used milk, because cow's, sheep's, goat's or camel's typologies undoubtedly justified different nutritional intakes in the final product. Also, the geographical and Country origin may have some importance (Abu-Lehia 1988). In fact, there are some chemical data available for sun-dried, freeze-dried, and spray-dried Jordanian *jameed*, showing clearly that sun-dried and freeze-dried products may reach 85–87.6% of dry matter, 9.05–10.9% of fat matter, and 14.9–17.9% of ash content (Al-Saed et al. 2012). On the other side, Jordanian spray-dried *jameed* shows 94% of dry matter, 8.2% of fat matter, an a reduced amount of ash (13.8%) (Mazahreh et al. 2008). However, some Syrian types may exhibit approximately 90.0% of dry matter and 5.5–5.7% of fat matter, demonstrating that certain regional/national difference may exist (Hasbani 2010).

4.3.2 Microbiological Profiles

From the microbial viewpoint, homemade and traditional *jameed* products are reported to have several problems when speaking of total viable count (TVC), coliforms, and *Staphylococcus aureus*. It has been reported that these values may reach 4.3, 2.0, and 1.6 Log_{10} (CFU/g), respectively, while *Salmonella* spp. contamination is demonstrated until 10 CFU/g. Consequently, the application of Hazard

Analysis and Critical Control Points (HACCP) approach to the production can reduce notably these amounts (Al-Saed et al. 2012). For example, TVC may arrive to 2.5 Log_{10} (CFU/g) while *Salmonella* spp. may be found as 'absent'. Interestingly, these results can be achieved even if the traditional process is not modified, differently from other above-reported studies (example: solar drying is carried out traditionally), while the control of fresh milk conditions is a critical step (Al-Saed et al. 2012).

4.4 *Jameed* and Traceability: The Flow of Input and Output Data

In the ambit of food traceability, the flow of input and output data should be based on a defined recipe. Because of the importance of *jameed* in Jordan at least, the origin of basic ingredients and minor additives should be recognised and justified (Sect. 4.3). Sensorial and technologically relevant properties of *jameed* (hardness, low moisture, salty taste, typical colour, etc.) should be maintained during all the commercial life of the product. This aspect is particularly important because *jameed* is normally used as an ingredient for the Jordanian *mansaf*; consequently, the durability has to be stated and be verifiable. Anyway, the basic flow of input and output data concerning traditional and industrial *jameed* in Jordan should concern only a simple list of ingredients. In the ambit of this book, we have decided to discuss the traditional recipe with one modification: the addition of a selected LAB culture.

4.4.1 The Traditional (Solar-Dried) Jameed

As explained in Sect. 4.3, the list of ingredients for traditional *jameed* (Fig. 3.1) should be summarised as follows (Al Omari et al. 2008):

a. Crude (untreated) sheep's or goat's milk (alternative possibilities: cow's milk, camel's milk)
b. Selected lactic acid bacteria (LAB) as yogurt starter culture. Similarly to *labaneh*, a *L. delbrueckii* culture can be considered (Sect. 3.3.1)
c. Salt.

In this ambit, the use of other ingredients (such as carrageenan) is excluded. As a consequence, the flow of input data (Fig. 4.2) concerns only three ingredients as expression of input data, while output data should concern only: the final *jameed* and the by-product (or OL), represented by separated butter. Each raw ingredient 'should' be considered as originated from one single location only, or by a single producer only. On the other hand, the Jordan economy shows a situation while a notable amount of collected (crude) raw milk is from many small farmers, including households and women's cooperatives (Al Hiary et al. 2013). As a result, the following critical points

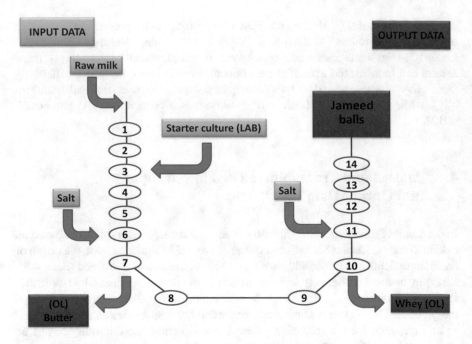

Fig. 4.2 Flow of input and output data for traditional *jameed*. The flow-chart implies several steps as follows, with some modifications if compared with the traditional method of the Bedouins (Al-Saed et al. 2012; Quasem et al. 2009): (1) heat treatment; (2) cooling; (3) addition; (4) incubation; (5) cooling; (6) addition; (7) churning; (8) heating; (9) cooling; (10) straining; (11) addition; (12) ball-shaping; (13) drying; (14) final packaging

may be observed, in spite of the really short list of used raw materials (and similarly to the management of *labaneh* traceability, as shown in Sect. 3.3.1):

1. The type of milk has to be considered (sheep—goat—cow—camel).
2. The identification of the milk collector, as always observed worldwide when speaking of milk cooperatives, concerns many small subjects.
3. The identification of the milk processor (*jameed* producer) is not required generally when speaking of low-scale productions, at the farm level (Al Hiary et al. 2013). A large-scale producer is responsible for the collection of physical raw milk by many milk collectors.
4. The identification of salt should be simple enough (one producer only), although medium-sized industrial plants may use different salt brands and types, with the consequent overload of traceability information. Also, salt may be added two times in the process (Fig. 4.2).
5. Similarly to salt, the identification of LAB starter culture should be simple enough (one producer only), although medium-sized industrial plants may use different LAB brands and types, with the consequent overload of traceability information.
6. The destination of the OL by-product (butter) should be clarified and always verifiable, being part of the flow of output data.

Figure 4.2 shows the situation by means of an industrial process. Differently from the traditional method, the example implies that raw milk is heat-treated, cooled, and the following steps include the addition of starter culture, a subsequent cooling, and other steps, as also reported recently (Al-Saed et al. 2012; Quasem et al. 2009). As easily explained, the 'traditional' recipe may imply a fixed number of ingredients (3), but the amount of collectors may be complex enough…

In addition, the problem of durability and also of the storage of traceability data should be considered, with specific relation to the duration of drying processes. In other words, an intermediate food which is seasoned only for a few days has a limited complexity when speaking of traceability data which have to be inserted and monitored day by day. The complexity is enhanced if the same intermediate has to be seasoned (dried) up to 10–15 days because of the coexistence of different raw materials, intermediates, and final products in the same plant. Normally, intermediates are segregated in dedicated rooms without possible contamination between different production flows belonging to different lots. On the other side, the same segregation has to be assured when speaking of traceability data which represent virtually the physical flow of production processes. Differently from real foods, information data may be confused and 'shared' or subdivided into different production data, without intention. This danger is real, and it normally increases when the duration of the total process becomes notable, similarly to the case of *jameed*. As a single example, *labaneh* is more manageable than *jameed*… because of the limited duration of the whole process. Duration is a critical factor of traceability networks, and great attention should be paid in this ambit.

References

Abu-Lehia IH (1988) The chemical composition of jameed cheese. Ecol Food Nutr 20(3):231–239. https://doi.org/10.1080/03670244.1988.9991003

Al Hiary M, Yigezu YA, Rischkowsky B, El-Dine Hilali M, Shdeifat B (2013) Enhancing the dairy processing skills and market access of rural women in Jordan. International Center for Agricultural Research in the Dry Areas (ICARDA) Working Paper No. 565-2016-38922, pp 1–12

Al Omari A, Quasem M, Mazahreh A (2008) Microbiological analysis of solar and freeze-dried jameed produced from cow and sheep milk with the addition of carrageenan mix to the jameed paste. Pak J Nutr 7(6):726–729

Al-Qudah YH, Tawalbeh YH (2011) Influence of production area and type of milk on chemical composition of jameed in Jordan. J Rad Res Appl Sci 4(4)(B):1263–1270

Al-Saed AK, Al-Groum RM, Al-Dabbas MM (2012) Implementation of hazard analysis critical control point in jameed production. Food Sci Technol Int 18(3):229–239. https://doi.org/10.1177/1082013211427783

Alu'datt MH, Rababah T, Alhamad MN, Obaidat MM, Gammoh S, Ereifej K, Al-Ismail K, Althnaibat RM, Kubow S (2016) Evaluation of different drying techniques on the nutritional and biofunctional properties of a traditional fermented sheep milk product. Food Chem 190:436–441. https://doi.org/10.1016/j.foodchem.2015.05.118

Barone M, Pellerito A (2020) In: Sicilian Street foods and chemistry—the Palermo case study. Springer International Publishing, Cham. https://doi.org/10.1007/978-3-030-55736-2

Bishouty JI (2000) The nutritive value and safety of some selected street foods in Amman. Dissertation, University of Jordan, Amman

Buhr BL (2003) Traceability and information technology in the meat supply chain implications for firm organization and market structure. J Food Distrib Res 34(3):13–26. https://doi.org/10.22004/ag.econ.27057

Carod Royo M, Sánchez Paniagua L (2015) Estudio del efecto de aditivos en la calidad de un snack a base de labneh. Facultad de Veterinaria, Universidad Zaragoza. Available https://zaguan.unizar.es/record/37002/files/TAZ-TFG-2015-3990.pdf. Accessed 30 Sept 2020

Cheek P (2006) Factors impacting the acceptance of traceability in the food supply chain in the United States of America. Rev Sci Tech Off Int Epiz 25(1):313–319

Chrysochou P, Chryssochoidis G, Kehagia O (2009) Traceability information carriers. The technology backgrounds and consumers' perceptions of the technological solutions. Appet 53(3):322–331. https://doi.org/10.1016/j.appet.2009.07.011

Fuquay JW, Fox PF, McSweeney PLH (2011) Milk lipids. Encyclopedia of dairy sciences, vol 3, 2nd edn. Academic Press, Oxford, pp 649–740

Giraud G, Halawany R (2006) Consumers' perception of food traceability in Europe. Proceedings of the 98th EAAE Seminar, marketing dynamics within the global trading system, Chania, Greece

Golan E, Krissoff B, Kuchler F (2002) Traceability for food marketing and food safety: what's the next step? Agric Outlook 288:21–25

Golan E, Krissoff B, Kuchler F (2004) Food traceability one ingredient in safe efficient food supply. Amber Waves 2(2):14–21

Hamad MN, Ismail MM, El-Menawy RK (2016) Chemical, rheological, microbial and microstructural characteristics of jameed made from sheep, goat and cow buttermilk or skim milk. Am J Food Sci Nutr Res 3(4):46–55

Hamad MNE, Ismail MM, El-Menawy RK (2017) Impact of innovative formson the chemical composition and rheological properties of jameed. J Nutr Health Food Eng 6(1):00189. https://doi.org/10.15406/jnhfe.2017.06.00189

Hasbani M (2010) Evaluation study of Syrian jameed in order to develop it. Dissertation, Damascus University, Damascus

Hilali M, El-Mayda E, Rischkowsky B (2011) Characteristics and utilization of sheep and goat milk in the Middle East. Small Rumint Res 101(1–3):92–101. https://doi.org/10.1016/j.smallrumres.2011.09.029

Iñiguez L, Aw-Hassan A (2005) 1.3. The sheep and goat dairy sectors in Mediterranean West Asia and North Africa (WANA). Special Issue of the International Dairy Federation 0501/Part 1, Proceedings of the Symposium on the future of the sheep and goat dairy sectors, Zaragoza, 28–30 October 2004, pp 13–22

Ishnaiwer M, Al-Razem F (2013) Isolation and characterization of bacteriophages from Laban jameed. Food Nutr Sci 4:56–66. https://doi.org/10.4236/fns.2013.411A008

Ismail MM, Farid Hamad MNE, El-Menawy RK (2017) Improvement of chemical properties of jameed by fortification with whey protein. J Nutr Health Food Sci 6(1):1–11. https://doi.org/10.15226/jnhfs.2018.001117

Jasnen-Vullers MH, van Dorp CA, Beulens AJM (2003) Managing traceability information in manufacture. Int J Inf Manag 23(5):395–413. https://doi.org/10.1016/S0268-4012(03)00066-5

JSI (1997) JS 462 (1997): Milk and milk products (Al Jameed). Jordanian Standard Institution, Amman

Mazahreh AS, Al-Shawabkeh AF, Quasem JM (2008) Evaluation of the chemical and sensory attributes of solar and freeze-dried jameed produced from cow and sheep milk with the addition of carrageenan mix to the Jameed paste. Am J Agric Biol 3(3):627–632

Quasem JM, Mazahreh AS, Afaneh IA, Al Omari A (2009) Solubility of solar dried jameed. Pak J Nutr 8(2):134–138

Shaker RR, Jumah RY, Tashtoush B (2003) Manufacturing of jameed using spray drying process. Int J Dairy Technol 52(3):77–80. https://doi.org/10.1111/j.1471-0307.1999.tb02077.x

Sodano V, Verneau F (2004) Traceability and food safety: public choice and private incentives, quality assurance, risk management and environmental control in agriculture and food supply networks. Proceedings of the 82nd Seminar of the European Association of Agricultural Economists (EAAE), Bonn, Germany, 14–16 May, Volumes A and B

Starbird SA, Amanor-Boadu V (2006) Do inspection and traceability provide incentives for food safety? J Agric Res Econ 31(1):14–26

Tawalbeh Y, Ajo R, Al-Udatt M, Gammoh S, Maghaydah S, Al-Qudah Y, Al-Sunnaq A, Al-Natour F (2014) Investigation of the antimicrobial preservatives in the dairy product (labneh). Food Sci Qual Manag 31:117–122

Verbeke W, Frewer LJ, Scholderer J, De Brabander HF (2007) Why consumers behave as they do with respect to food safety and risk information. Anal Chim Acta 586(1–2):2–7. https://doi.org/10.1016/j.aca.2006.07.065

Printed in the United States
By Bookmasters